T0313372

Managing People at Work
A New Paradigm for the 21st Century

RIVER PUBLISHERS SERIES IN MANAGEMENT SCIENCES AND ENGINEERING

Series Editors:

J. PAULO DAVIM
University of Aveiro
Portugal

CAROLINA MACHADO
University of Minho
Portugal

Indexing: All books published in this series are submitted to the Web of Science Book Citation Index (BkCI), to SCOPUS, to CrossRef and to Google Scholar for evaluation and indexing.

The "River Publishers Series in Management Sciences and Engineering" looks to publish high quality books on management sciences and engineering. Providing discussion and the exchange of information on principles, strategies, models, techniques, methodologies and applications of management sciences and engineering in the field of industry, commerce and services, it aims to communicate the latest developments and thinking on the management subject world-wide. It seeks to link management sciences and engineering disciplines to promote sustainable development, highlighting cultural and geographic diversity in studies of human resource management and engineering and uses that have a special impact on organizational communications, change processes and work practices, reflecting the diversity of societal and infrastructural conditions.

The main aim of this book series is to provide channel of communication to disseminate knowledge between academics/researchers and managers. This series can serve as a useful reference for academics, researchers, managers, engineers, and other professionals in related matters with management sciences and engineering.

Books published in the series include research monographs, edited volumes, handbooks and text books. The books provide professionals, researchers, educators, and advanced students in the field with an invaluable insight into the latest research and developments.

Topics covered in the series include, but are by no means restricted to the following:

- Human Resources Management
- Culture and Organisational Behaviour
- Higher Education for Sustainability
- SME Management
- Strategic Management
- Entrepreneurship and Business Strategy
- Interdisciplinary Management
- Management and Engineering Education
- Knowledge Management
- Operations Strategy and Planning
- Sustainable Management and Engineering
- Production and Industrial Engineering
- Materials and Manufacturing Processes
- Manufacturing Engineering
- Interdisciplinary Engineering

For a list of other books in this series, visit www.riverpublishers.com

Managing People at Work
A New Paradigm for the 21st Century

Murali Chemuturi

Chemuturi Consultants, USA

Vijay Chemuturi, MBA, CPA, CFE

USA

Routledge
Taylor & Francis Group

LONDON AND NEW YORK

Published 2019 by River Publishers
River Publishers
Alsbjergvej 10, 9260 Gistrup, Denmark
www.riverpublishers.com

Distributed exclusively by Routledge
4 Park Square, Milton Park, Abingdon, Oxon OX14 4RN
605 Third Avenue, New York, NY 10017, USA

Managing People at Work A New Paradigm for the 21st Century / by Murali
Chemuturi, Vijay Chemuturi.

Routledge is an imprint of the Taylor & Francis Group, an informa
business

ISBN 978-87-70221-08-5 (print)

While every effort is made to provide dependable information, the
publisher, authors, and editors cannot be held responsible for any errors
or omissions.

Contents

Preface

Why is a book on human resources management necessary? Are there no adequate number of books already available on this subject in the market authored by experts in academia? You are right to ask these questions and we would be surprised if you don't!

Most non-fiction books are authored by the dons from the academia. They conduct research on a relevant topic and gather data from various sources and subject them to analysis. Then, the findings from that research are collated into a book and then published after a peer review. These books are used as text books or prescribed reading in the university courses. Rarely do we find books authored by practitioners from the industry.

Academicians gather data from the industry which obviously are second hand in nature. Unless the data are gathered using first-hand observation, the veracity of such data is questionable as the insiders in the industry sanitize the data before giving them out. When the data are dubious, can the results obtained by analyzing such data be accurate?

In physical sciences like physics, chemistry and their variants, academia leads the industry as industry does not have the luxury of conducting experiments just to prove a point. But in social sciences like management, marketing and economics, the industry leads the academia since they have no other alternative to experimentation to profitably manage their organizations. In these fields, the academia can only gather past data or what happened and the past data are not a very reliable indicator to predict what would happen as the environment keeps on metamorphosing itself continuously.

The scenario of organizational management went through a metamorphosis. The days of "making a pin to a locomotive engine within the organization" simply obsoleted! Now it is "we outsource/offshore most of our deliverable" is the name of the game. The famous dictum of Jack Welch was, "70-70-70" which translates to 70% of what we market is out sourced; 70% of what we outsource, we offshore; and then 70% of what we offshore would be to organizations dedicated to our work. This changed the game. While outsourcing was there for a long time, it was on a much lower scale. It was referred to as sub-contracting and was resorted to as a measure to supplement our capacity or to take advantage of a facility which was not cost-effective to setup in-house. Now, the large multinational organizations do just marketing and, in some cases, designing and the labor-intensive manufacturing or development is outsourced!

In the recent past, most of the communication and information-processing work was carried out on paper. Computers were restricted to EDP rooms limited to processing bulk data. Record-keeping was achieved by filling up the registers and filing cabinets. Stenos and typists were doing the transcription work. Now, we do not think any workstation, in any department, exists without a laptop/desktop as the main tool for doing the work. Registers, files, papers (other than printouts), pens, pencils, clips and pins simply did the vanishing trick from the offices. Employees with designations like secretaries, assistants, typists, clerks cannot be found in the present-day organizations. The layered hierarchy gave way to flat organizations. The levels in the organization dropped from nine or ten to three or four.

In this environment, management is not simply getting things done any more! It is rather "managing the results" expected of the position. Very few people in the organization are designated as managers as not many have more than five reportees! Managers work with their own hands using the decision support tools. Now there is no system of "managers" and the "managed" in the organizations.

Just as every field of management, including the financial management, marketing management, material management and operations management, metamorphosed, the management of the organizational human resources also went through a metamorphosis. While the industry moved forward, the academia and the books that record the contemporary knowledge stayed behind. It is our intention to bridge this gap and present the state of the art in the field of human resources management to our readers.

We love to hear your feedback. Please feel free to email us murali.chemuturi@gmail.com and we promise to answer every email we receive, mostly within a business day or at the earliest.

List of Figures

List of Tables

1

Introduction to People Management

1.1 Introduction

In our opinion, people management is the oldest form of management. The word "management" itself contains the word "man", representing the human beings right at the beginning of the word. We manage people for achieving the objective of completing the deliverable meeting the goals of being on time, within the allocated budget and at the same time maintaining and improving the morale of the team members. As part of people management, we carry out the following activities:

1. Estimate the requirement for people.
2. Acquire the required human resources to perform the activities.
3. Allocate work to team members such that the execution is efficient and effective.
4. Motivate the team members towards higher levels of achievement.
5. Maintain the team morale at levels desired by the organization.
6. Discipline the people if necessary.
7. Develop and mentor suitable persons for shouldering higher responsibilities.

Let us look at each of these activities in greater detail.

1.2 Estimate the Requirement for People

We need to estimate the requirement for people to perform the activities allocated to us. This would be based on the estimated workload and the productivity baselines of the organization. Each operation needs a specific skill set. People also come with multiple skill sets and can perform multiple operations. If an operation provides work load continuously for all the 8 hours of the day and day after day, then we designate a person to carry out that operation. If not, we combine two or more operations that would need a similar skill set and designate a single person to perform all of them. You can see a good example of this in the receptionist. Function of a receptionist is basically to receive the visitors and help them. But as the work load is insufficient to engage all her time, we usually entrust her with receiving phone calls, mailing work and so on.

In this manner, we designate positions for our department ensuring that each of the individuals employed shall have work for all the working hours day after day. Then for each position, we record the job description listing all the activities expected of the position and the desired skill set. We need to recognize the reality that it is rare that any position would have just the right amount of work. For some, it fills all the working time; for some a little spare time is left; and for some, it needs slightly more than the available time. Here we need to note that our estimation is assuming an average level of skill. In such cases, we have a few alternatives such as the following:

1. When a position is projected to have some spare time, it is better to employ a person with lower skill level than the assumed level. Skill levels are discussed in the chapter on work management in detail. That way, we can pay a lower rate and also ensure that the individual has full workload.
2. The other alternative available for the position with spare time is to assign some other work like preventive maintenance and so on.
3. Alternatively, allow the person to enjoy the spare time made available by the position.
4. When the work needs slightly more than the available time, one alternative is to employ a person with a higher level of skill than the assumed skill level.
5. An alternative is to allow the individual to work harder initially for some time during which the skill level of the person would improve so that all the work can be performed in the available time itself. Human beings are capable of raising their skill levels almost without limits. But this alternative, while viable, may affect the motivation of the employee and even health too.
6. Another alternative to overload is to pay extra money as incentive or overtime pay so the individual would not lose motivation.

Once we estimated the number of individuals needed, their skill sets and the skill levels, we recruit them in coordination with the HR department and induct them.

1.3 Acquire the Required Human Resources

People can be sourced either from within the organization or from outside. We may have people within our organization, especially, if it is a running organization, available for transfer.

1. The organization may have other departments which may have spare capacity or reduced workload and the people working in that department may be available.
2. The organization may decide to close some department and may be laying off people.
3. Sometimes people may express to or be required to be shifted from their positions for various reasons including change of geographic location to be near their families, or health reasons or any other reasons.
4. The organization might have acquired another company and people from that organization might become available.

It is possible, for the above reasons or other reasons, that people may be available from internal sources. It is beneficial to retain talent than to recruit from outside. It may need some retraining or reorientation to make them suitable for the position. So, when available, we can recruit the required people from internal sources.

Sources external to the organization include:

1. There are organizations that supply employees for working with other companies. We can hire such people for some time with a stipulation that we can absorb them on to our rolls. This has the advantage that the individual need not be fired if (s)he is found to be a misfit. We can just repatriate that person.
2. There are many freelancers out there and they can be hired.
3. New recruitment for the regular rolls of the company.

It is the responsibility of the HR department to recruit the required number of people with the desired skill sets. We need to interact with the HR department and acquire the required human resources.

1.4 Induct the Acquired People

Once human resources are sourced and are allocated to us, we need to induct them into our department. We may or may not conduct formal induction training, but we do provide the following information while inducting them.

1. The responsibilities of the position
2. The methods of carrying out the allocated work
3. The methods of management and oversight
4. The methods of quality assurance
5. The procedure for issue resolution
6. The procedures for grievance resolution
7. The methods for material and wastage handling
8. Any other work-related instructions

Once all this information is conveyed to the employee, (s)he is ready to shoulder the responsibilities of the position. (S)he is inducted into the workforce of the department and work is allocated to that person for execution.

1.5 Motivate the Team Members Towards Higher Levels of Achievement

When people come to work, they typically have two primary objectives, namely, to earn a livelihood for them and their loved ones by performing the work assigned to be able to receive their paycheck and learn or develop their skill sets that can be traded in the future. Then there could be a host of secondary objectives like possibility of career advancement and earning more money, pride and satisfaction from doing work, respect in the society and others. Regardless of secondary objectives, the first objective is to earn a livelihood!

There are, of course, exceptions, which only prove the rule. Take the example of people whose livelihood is already taken care of by their fore-fathers or spouses, still they might come to work. In these cases, the primary objectives might include personal and/or self-respect that comes with working, acquire knowledge for instance to learn about the business in order to start a similar one, and social interaction available at the workplace to pass the time productively.

The work is generally a means to an end rather than the goal itself. We can classify people at work in a variety of ways.

A simple classification of people (by purposefulness) is:

1. Straightforward (open) people
2. Scheming (closed) people
3. Normally straightforward except when their interests are affected

Yet another way of classifying people (pleasantness) is:

1. Pleasant people
2. Unpleasant people
3. Neutral (to pleasantness or unpleasantness) people

Still one more way of classifying people (expectations) is:

1. Easy to please people
2. Difficult to please people
3. Normally easy but could be difficult on occasion

One more way of classifying people (ambition) is:

1. Ambitious people
2. Un-ambitious people
3. Somewhat ambitious

One more way of classifying people (willingness to accept you) is:

1. Willing people
2. Reluctant people
3. Normally willing (except when you bear bad news) people

One more way (responsibility) is:

1. Responsible (in nature) people
2. Irresponsible (in nature) people
3. Mostly Responsible

To conclude classification with this last (age) class: (note that is an illegal means of classifying individuals in North America and Europe at a minimum)

1. Those in the prime of age
2. Those well past their prime
3. Those slightly past their prime

In all of these myriad classification schemes, choices one and two are the extremes while the third choice is typical of people in organizations.

An individual can be any combination of the above classifications. For example a person can be –

1. Straightforward, pleasant, easy to please, ambitious, willing, responsible, and in the prime of age or
2. Scheming, unpleasant, difficult to please, un-ambitious, reluctant, irresponsible, and well past prime.

The list can and does go on a multitude of combinations and permutations. Given this level of complexity, how do we motivate a person to perform their task to the best of their ability?

History is replete with examples of ordinary people performing extraordinary feats when motivated by necessity, fear or reward. Motivation techniques are dealt in greater detail in the chapter on morale management within this book.

Team morale is extremely important if the work is to be executed effectively. Even underdogs can be exhorted to incredible actions if their morale is very high. We often see in sports the team with higher morale always wins over the team with lower morale even if they have similar skills. Even if every team member is positive the team morale can be lower. Some of the reasons for lower team morale could be:

1. Perception that the work is not important to the higher management in the organization.
2. Perception that the technology used is obsolete.
3. Perception that the human resources are being treated unfairly by the management.
4. Perception that the team meetings and reviews are not receiving adequate importance.
5. View that the targets are either too lax or too tight.
6. Presence of rumor mongers and gossips in the team.

There could be many other reasons that can lower morale. What is important is the manager should continuously monitor the team morale and strive to maintain it at desirable levels. Motivation and morale are dealt in greater detail separately in the chapter on morale management.

1.6 Discipline the People as Needed

While we take every precaution and action necessary to motivate people and maintain the team morale at a high level, sometimes it becomes necessary to discipline erring resources. Disciplining becomes necessary only when an employee is found to be willfully indulging in acts that are detrimental to the health of the operations/project and morale of the team. Unintentional delays to the assigned tasks or taking more effort than estimated may necessitate counseling and coaching but not disciplining. People acting out may also be a result of unresolved grievances. Before we subject an employee to disciplinary actions, we need to ensure that:

1. There are no unresolved grievances.
2. Some of the reasonable expectations of the employee have not been met by the organization.
3. The misdemeanor is not the result of an innocent action.
4. The offending act is indeed willful.
5. There is no conspiracy against the employee by others.

Once the above criteria have been applied, we may take a disciplinary action ensuring that a mistake is not being committed. One cardinal rule to remember is that "the punishment ought to be commensurate with the misdemeanor". Termination of the services of the employee is akin to capital punishment. Recruiting a new employee and training the person costs much more than correcting and retaining an existing employee. The result of disciplinary action ought to be targeted at correcting the behavior rather than inflicting a punishment. In most cases the misdemeanor probably does not crop up overnight. Behavioral issues tend to be the culmination of number of incidents over a period of time. We typically find that discipline is a failure of management that can be prevented through proper motivation and behavioral correction.

When discipline is inevitable, we suggest that managers follow the disciplinary procedure of the organization diligently giving every opportunity to the erring employee to redeem him or herself.

1.7 Develop and Mentor People for Shouldering Higher Responsibilities

Managers are uniquely positioned in the organization for spotting talent for leadership in the juniors. Once leadership potential is discovered in an employee the manager is usually the first person in the organization that can develop the potential into reality by putting the latent talent to practical use. By getting the person to exercise the latent leadership talent the manager creates an opportunity for the employee to develop. Mentoring involves making the employee part of decision-making process, providing leadership opportunities beginning in a small team environment, continuously coaching the person and then increasing independence gradually over a period of time. We also suggest exposing the new leader to senior levels of management so that (s)he can be evaluated as well as to continue the person's development by enhancing the opportunity for the new leader to be allocated to other positions in order to work with other managers within the organization.

1.8 Performance Appraisals

We need to appraise the performance of the employees periodically. Every organization carries out formal periodic performance appraisals at least once a year and some organizations do have more frequent formal performance appraisals. But we need to carry out informal performance appraisals in a regular manner. Informal performance appraisals provide an early warning to the poorly performing employees to bring their performance on par with other employees and allow them to get a better rating in the formal performance appraisals.

Performance appraisals are a very important aspect as the pay hikes, rewards and career advancements are based on the formal performance appraisals in most organizations. For the individual employees the performance appraisal is the recognition of their efforts and contributions and constructive feedback on areas requiring improvement in their performance. When an employee is rated positively, there would be no concerns but if the rating is not so positive, there would be a lot of concerns. If the process is not handled tactfully, giving a poor rating can result in bitterness and lead to deterioration of morale not only for the individual but for the whole team. While the performance appraisal is confidential, it is common among employees to discuss the ratings among themselves in an informal manner. So, it is not just absolute rating but also about the relative rating that the employees fret about.

Therefore, when assigning ratings to the employees, it is essential to base the ratings on objective criteria. We can use objective criteria only when we maintain meticulous records of the performance. This record would come in handy to convince a complaining employee about the fairness of his/her rating. Besides, the records come in handy to defend yourself when the aggrieved employee escalates the issue to higher management. Human beings are very tricky and some of them are not straight forward. They raise a variety of issues when they feel that they were not given the rating they deserve. Some of these issues could be:

1. That the manager is a bully.
2. That the manager is biased racially/religion/language/nationality or in any other manner.
3. The manager is sexually harassing.
4. Other unreasonable issues.

These issues could be raised behind your back. So, be sensitive, be very careful and maintain meticulous records while carrying out formal performance appraisals. When you need to give poor rating to a junior, take your boss into confidence. It is also wise to take the HR representative into confidence in such cases.

Performance appraisals are very important and are used by the organizations to select people for promotions and special assignments. So, we must make diligent efforts to give appropriate rating to each of our juniors based on meticulously maintained records. That way, we are giving the organization an opportunity to utilize the formal performance appraisals to be used in decision making.

1.9 Release of the Employee

We may need to release an employee for a variety of reasons including transfers, resignations, terminations, layoffs and so on. We need to release an employee in such a way that the loss does not impact the functioning of our department/project. We perform the following activities in releasing an employee to ensure that the organization is not put to loss:

1. In the case of transfers, carry out the performance appraisal for the employee and after agreement with the employee, handover the performance appraisal to the HR department.
2. Update the skill database of the organization with the new skills acquired by the employee under your supervision.
3. Takeover all official artifacts including documents, drawings, tools and any other artifacts that may be with the person.
4. Archive the official communications that may need preservation for future reference and possible use.
5. Prepare the release communication (a formal letter or an email) with guidance from HR department and refer the employee to the exit interview process.
6. Document best practices, worst practices or any events of the individual deserving special mention and forward to concerned management personnel.
7. Update the concerned records with information to reflect the release of the employee.

This completes the release of the employee. Sometimes we may employ temporary hires or contractors. When we release them, the process would be similar to the one described above. Overall, the release of an employee needs to be carried out in such a manner that the organization or the department should not be adversely affected due to such release.

1.10 Some Best Practices in People Management

Human resources have to be treated with a different approach than you treat other resources. Once you are out of the organization, the machines, the materials or the money belonging to the organization do not follow you or maintain contact with you. But human resources on the other hand would continue to be in touch with you and may even follow you or in some cases precede you to another organization. It is possible that your junior in one company can be your junior, peer, or even your boss in a new organization.

So, it is better for you to treat your juniors as you would treat your own children. Why did we say children? You know that your children would err and you would forgive them; they would commit mischief and you would correct the situation; they would go out and fight with others and you would console them; they get into trouble and you extricate them; they would not show interest in their curricular activities and you coach them; the list is endless and you still do all that without expecting anything in return. Their growth and appreciation of your efforts is all the reward you want. Juniors are like that, demanding continuously from you. In return, they do the work assigned to them for which they are paid. If you ill-treat your children, they tell the whole world and so are the juniors. Your children would ultimately grow up and realize your value and their follies but juniors may not.

The following suggestions will help you manage your human resources effectively.

1. Communicate continuously and lucidly with all team members. In the absence of official communication, employees lend their ears to organizational grapevine which can distort facts in a way that is detrimental to the department's wellbeing. There is a separate chapter on managing communication effectively in this book.
2. Monitor morale continuously and strive to keep it at a high level. A team with high morale can scale unimaginable heights.
3. Motivate continuously to learn about the team members and try your best to motivate them. Motivation to some extent is individual and it pays to show individual attention.
4. Discipline only when you are absolutely sure it is the only available alternative. Remember that unless you administered discipline very skillfully your actions can de-motivate the entire team. Make every effort to be not only fair but also seen to be fair to the entire team.

1.11 Handling Difficult People

As a manager, you need to handle various people ranging from your team members, peers, superiors, vendors and customers. One thing typical for a manager is that all these persons are highly intelligent. Most of the people are good and perform what is expected of them. But few are difficult. Would a few difficult people really matter? George Bernard Shaw once said, *"Most people in the world adjust them to the ways of the world. Only a few would want the world to adjust to their ways. Therefore all progress depends on them"*. Remember the dictum "Insignificant majority and significant minority" – this is used in many situation like ABC (Always Better Control) Analysis, VED (Vital, Essential and Desirable) analysis and so on. It is one Lord Krishna, one Jesus Christ, one Mohammed, one Lenin, one Hitler, one Mandela, one Martin Luther King and one Gandhi that changed the world. Hence, here is an attempt to forewarn you about this significant minority so that you can be prepared to effectively handle all your people.

We are reminded of an anecdote. A psychiatry Professor was giving lecture to the students about the behavior of the abnormal man – for five days in a row. When the Professor started to continue the lecture on the sixth day, a student got up and asked "Professor, when will you begin telling us about the behavior of the normal man?" "When we find him", the professor replied, "We cure him".

Are there any people that are easy to handle? Some people may be, but for only some time. Who are these people, easy to handle for some time?

1. Trainees, till their training is completed and are confirmed on the permanent rolls of the company.
2. New employees, till their probation period is successfully completed.
3. Employees that are expecting a promotion and wishing to get it.
4. Employees that are low on performance but high on pleasing the boss and thus survive.
5. Eager-beavers, who always want to please the boss just for the sake of pleasing but their density in the employee population is very low.
6. Wise people understand their role as well as yours and perform their duties diligently irrespective of your stimulus and are always easy to handle. Their percentage is miniscule in the organizations.

The rest are, to a degree, difficult to handle. Here we attempt to profile the difficult people and give you some cues to handle them.

Why should we handle these people? Because we cannot avoid dealing with them. They are part of every organization and who knows, we ourselves could be difficult for others.

1.11.1 Classification of Difficult People

We can classify difficult people into the following categories –

1. Two-timer
2. Backstabber
3. First Chapter Expert
4. The Martyr
5. Prima Donna
6. Manipulator
7. The Gossip
8. Breather-down-the neck
9. Buck-stopper
10. No-man
11. Mr. Justice
12. The Courier Pigeon

Where do you find these people – in your juniors, your peers and in your superiors too.

1.11.2 Two-timer

Here is a person whose stand depends on the situation and the people involved. He wears "public smile and a private snarl". What he says in public differs from what he says in private. This person is completely undependable.

We can spot him the first time we are two-timed. Do not treat the first one as coincidence. No one two-times unintentionally. These people could be found in your juniors, peers as well as superiors.

How do we handle this person?

When he makes a private commitment to you, do not depend on it unless you get it in some form of writing such as email and if he chats on a messenger, save the conversation. Try to have witnesses when you deal with him. Maintain minutes of meeting whenever you meet him.

1.11.3 Backstabber

Backstabbers betray confidence, especially if it can get them some positive points with the upper echelon. They are not your enemies but could be your friends. Very difficult to spot these people or prove that they backstabbed you.

We can spot these persons only through secondary sources. When a boss-type person is exhibiting animosity toward you for no apparent reason, realize that somebody backstabbed you. And that somebody is suddenly friendly with this boss-type person. These people could be found in your peers and juniors and rarely in superiors.

Backstabbers obtain the information that assists them to backstab you, from you, using charm, encouragement and sympathy to encourage you on to criticize senior persons. Remember, "Loose lips sink ships". In your workplace, never offer criticism, in the presence of others, of a senior person who is especially powerful.

Backstabbers can backstab you only when your back is turned towards them. So never show your back to any one including the most charming friend of yours.

1.11.4 First Chapter Expert

This person knows something about every scenario, technology and person in the organization. He never allows anyone to get into details because his knowledge is limited to the introductory first chapter of a book on the subject at hand. He uses this limited knowledge to shoot down any positive proposal by picking holes, which are plugged in the later chapters of the book. It is extremely difficult to argue with a chap who has half-baked knowledge on the subject.

It is easy to spot this person. You can't miss him as he is often found discussing new topics all over the place and disappearing as soon as someone gets into detail. These people could be found mostly in your peers.

How to deal with this person? Get into the details first and give the guy some credit saying something like "As our John probably can explain to you/knows . . .". This normally shuts him up.

1.11.5 The Martyr

Often, people who are overlooked for promotions, become martyrs. Especially those people with a long tenure in the organization and whose academic credentials prevent them from reaching the top echelons or those persons that are not on the fast track of career progression, exhibit the attributes of a martyr. Martyrs give a negative spin on everything, especially in private. They do not do so in public as they fear getting fired from the organization.

It is easy to spot them. When something goes wrong they come to you and say, "I told you so". Whenever there is a new initiative, they would tell you in private that it is going to fail. They are harmless except that they discourage you from coming out with new initiatives or take up something challenging. These people could be found in your superiors and peers.

Indulge him, but never take his discouragement seriously. He is also a great source to point the other side of your proposals. Hence, use him as your quality assurance person for your proposals and initiatives.

1.11.6 Prima Donna

These are sticklers for rules, regulations, conventions, practices and so on. They are normally petty-minded people and could be found in administrative assistants, security department, secretaries to bosses, auditors and so on.

Have you ever had experience with the customer service staff at governmental agencies, such as the motor vehicles department, who will take his/her own sweet time to attend to your needs? He is a Prima Donna.

You visit a C-suite executive and his secretary tells you to wait. You see her polishing her nails but not making any attempt to communicate your arrival with her boss. You visit the staff in the court clerk's office and are asked to wait as the representative is chatting with her co-worker. You try to remind her and you get reprimanded. She is a Prima Donna.

These people have limited power, just to annoy you, and they use it to its full extent.

How do you handle them? Say "Hi" to the such people when entering the premises and prepare to be patient. Put on your charm and best manners with the secretary.

The trick is to give them an impression that you respect their position and power and you can get along well with them.

1.11.7 The Manipulator

Manipulators feign to be very busy to either avoid work or to pass it on to you. They always look haggard and under pressure. They are either slow workers or do something else during working hours. They come to you with something like this "I am very busy right now, can you help me and do this"?

Or they tell the boss that even though it is the manipulator's job, you are the best person in view of your past experience or special skills (real or imagined) or that you "seem" to be free of any work on hand.

These people could be found mostly in your peers.

When you detect this "always busy-as-a-bee" person, try and put on that haggardly look yourself and approach him to see if he is free to help you out. Are we telling you to become a manipulator yourself? You bet we are, there is no medicine that is better suited to a manipulator than a dose of his/her own medicine.

1.11.8 The Gossip

The problem with Gossips is that they pick up loud thinking or parts of conversation, then put two and two together to make it eight and pass it off as fact. If they do not find anything worth circulating, they often invent juicy information. They sow distrust, prejudice and suspicion among people. They also waste time – yours and theirs.

They normally have spare time and also are very good conversationalists. They narrate with great skill and hold your interest. They make you ask for more. They are addictive. These people could be found in your juniors and peers.

The best way to handle them is to avoid them or avoid being drawn into a gossip session. Remember, if they are spreading false information about someone else with you, they surely would do the same about you too, with others when you are not there.

1.11.9 Breather-down-the Neck

(S)he is an uneasy-delegator. (S)he is usually in the position of a boss and perhaps due to some past bad experiences with his/her juniors, is not comfortable delegating the work. The person is also insecure about his/her position. When a person who normally does wok by oneself gets promoted to the position of a boss, that person also becomes a breather-down-the-neck till (s)he attains maturity. Sometimes bosses with few juniors also resort to breathing down the necks of their juniors as they have nothing better to do.

Understand that (s)he is growing up and maturing to be a boss and give space. One thing that gets him/her off your back is to meet the deadline the first time and a few more times so that (s)he becomes confident of your commitment. Then that person would not be so overbearing. The trick is to give confidence about your commitments.

1.11.10 Buck-stopper

Anything that goes to the buck-stopper – stays there. It never comes out. If you want the buck to be moved, you need to chase the buck-stopper. This person never says no, nor gives a commitment nor does actually respond. His/her desk is a bottomless pit. You can put in anything but getting something out is very difficult.

This person generally dodges all your queries about committing a date for any action expected of him/her. You can find these people in service departments handling grievances or complaints. These people could also be found in your peers especially in the service departments.

Their philosophy is "If it is really necessary or urgent, someone will drop by in-person". Hence, to get things done by such persons, be informed that you need to push the buck. Pay a visit and get the person to move it.

1.11.11 No-man

These persons are capable of saying "No" to each and anything that you would say. They have excellent logic to deny your proposal or request. You say "The Sun rises in the east" and these persons say "No the Sun does not rise in the east". If you press on with your argument, they would argue that east itself is ill-defined, physical North Pole and magnetic North Pole are not the same, that earth itself is slanted and so on.

It is easy to spot them. The most frequent word spoken by them is either "no" or "not". These people could be found in your peers especially in service departments.

While so, these people would be ready to put forward proposals. So, if you want to get them to say "yes" or something similar, you have to make them put up your proposals on your behalf. Then they would be saying "no" to all those that criticize your proposal! Instead of telling them what needs to be done, consult them and give suggestions and make them perceive your proposal to be theirs.

1.11.12 Mr. Justice

These persons are brought up on the stories of class-struggle. They divide the world (organization) into haves (of power or management) and have-nots (staff, workers, professional workers and so on). They see injustice in every action of the management. They denigrate the benefits but always, accentuate the side effects.

You will more often than not find them active directly or indirectly, in trade unions and such other associations. You will also find them opposing recognition of merit in persons and advocate seniority for awards and rewards. Consequently, you would also find them low on performance. They would be hovering around penalty-avoidance level of performance themselves. These could be mostly found in your juniors and rarely in peers.

Confrontational-counseling, using quantitative data is the best way to bring them into line. And this is a continuous and periodic chore. They will revert to trade union jargon, if you miss one confrontational-counseling session.

1.11.13 The Carrier Pigeon

The bearer of good news always gets the reward. You would have seen this in movies; the heroin hugs and/or kisses the postman bringing a letter from her distant lover or from an agency informing her of some good news! So, some people in the organization take on this postman role. Wherever something gets done, these persons go and inform the boss before the actual performer has a chance to get to the boss. They always keep their antennae up, scanning the horizon for newsworthy items to carry to the boss. While this is harmless, your punch is lost. Your latest success is already old hat.

You can spot them by observing for those who poke their nose into other's affairs. When something important is going on in the organization, these persons would be hovering around even if they are not involved. You will find them at places where they are not needed. These people could be found in your peers mostly.

Obviously, these people have the ear of the boss. So be careful what you say to them. They not only carry good news but also carry tales. If you criticize the boss in this person's presence, be sure that it would be carried. Hence, discretion is your safeguard. Do not reveal if you are on the verge of some success. Keep your calm and show your excitement only in the presence of your boss.

1.11.14 Final Words on Handling Difficult People

Human beings are generally difficult to handle, as they are unpredictable. Their response is not commensurate with your stimulus, most of the times. People also change their personality attributes over a period of time. It is not necessary that the dictum "once a thief always a thief" is always correct. But "forewarned is forearmed" is better than ignorance.

As a manager, you ought to know how to motivate people towards success and practice it diligently. But it is also necessary to be prepared to handle difficult people. The purpose of this section is to "forewarn you and to forearm" you.

2

Organizational Framework

2.1 Introduction

Organizational framework is the platform of the organization on which all the employees can perform to their best in the achievement of the organizational objectives and goals. The main responsibility of putting the organizational framework in place rests on the shoulders of the top management. They may consult the middle managers to get the best possible thinking on the subject, but the onus rests with the top management. Not only that, keeping the organizational framework in its original working condition as well as improving it or changing it to suit the changing business environment also rests firmly on the shoulders of the top management. In the matter of people management in the organizations, the onus rests with the HRD (Human Resources Department) of the organization. It includes the following:

1. Definition of processes for people management in the organization. This would include:
   (a) Procedures for requisitioning people for performing the organizational activities
   (b) Guidelines for selecting the source of people, that is, temporary resources from outside, from inside the organization for a project; guidelines for bringing in freelance consultants; guidelines for selecting contract employees and freelance consultants and all such activities
   (c) Procedures for actions such as recruitment, promotions, separations, disciplinary actions, and so on
   (d) Various standards for the quality of people that can be employed in the organization, on the job performance, code of conduct and so on

    (e) Formats and templates for use in all the people related activities

    (f) Checklists to verify comprehensiveness of various artifacts produced in people related activities in the organization

2. Provide resources and funding for the performance of people related activities.
3. Institute a reward and discipline system to motivate and encourage the employees to excel.

Let us discuss each of these activities in detail in the following sections.

2.2 Championing the Organizational HR Processes

The HR department working at the organizational level defines a process for people management and implements it across the organization. Then, based on the process improvement requests received from various stakeholders, it improves the process periodically adhering to the procedure for process improvement to keep the defined process current and eliminate all entropy in the system. The following activities are performed by the HRD department to be the champion of the people related processes in the organization:

1. Arrange for the definition of the processes drawing upon the expertise from the practicing managers from within the organization. This includes giving facilities to the experts, provide resources like time off from their regular routine and so on.
2. Arrange for the quality assurance of the defined processes by independent people and compile the feedback and arrange for review and implementation of the feedback into the defined process.
3. Implement the defined process on a pilot basis and collect feedback from that implementation; review it and arrange for implementation of the selected feedback into the defined processes.
4. Arrange for the managerial review of the finalized processes.
5. Implement the feedback obtained from the managerial review.
6. Obtain approval of the finalized process from authorized functionaries of the organization.
7. Train key players in implementing the approved process and internalizing it across the organization.
8. Roll out the approved process across the organization and implement it.
9. Arrange periodic audit of the process implementation in the organization and resolve non-conformance reports, if any.

10. Initiate maintenance activities of the process which include:

   (a) Receiving process improvement requests from the periodic audit reports, from users of the process and from management personnel as and when they are raised
   (b) Reviewing all such process improvement requests to select the requests appropriate for implementing in the processes
   (c) Arrange for updating the existing process elements with the approved process improvement requests
   (d) Arrange for review of the updated process elements to ensure quality and collect review feedback
   (e) Implement the review feedback in the updated process elements
   (f) Arrange for managerial review of the updated process elements by the approving authority and arrange for implementation of such feedback in the updated process elements
   (g) Obtain approval from concerned functionaries for the updated process elements
   (h) Release the updated process elements and collect back the obsolete process elements so that the outdated elements are no longer used in the organization
   (i) This is carried out periodically

In addition to these activities, HRD generally promotes implementation and conformance to the defined process and process driven working in the organization. HRD would also be the front end in the certification of the organization by participating in the certification audits defending the people management practices in the organization.

2.3 Funding

Funding the business operations is the responsibility of the entrepreneurs. The top management is the representative of the investors. The funding needed by the organization is of two types: fixed capital and the working capital. Fixed capital is the investment in fixed assets such as buildings, machinery, tools, furniture and fixtures, and so on. Working capital is the money we expend to run the business operations. Working capital is utilized to procure items that are converted into deliverables and sold. Salaries and consumables like stationary, food items, and so on also form part of working capital. Fixed capital is required initially at the time of setting up the facilities for the organization as well as at the time of expansion or for maintenance and replacement of the facilities. Working capital is required regularly for meeting the expenses of running the organization as well as for building our deliverables. Marketing expenses also fall under working capital. All salaries form part of working capital. We do not wish to make this a treatise on the financing principles here and we just wish to give an idea of the differences between fixed capital and working capital.

The top management needs to set in place a system for predicting the requirements of capital, both fixed and working, evaluate the requirements, approve, reject or postpone the requirements. This system is referred to as budgeting and is usually carried out by the finance department. The procedures for raising fund requirements, their evaluation, recommending for approval, approval, allocation of funds, utilization of funds, reconciliation, and checks and balances thereof are the responsibility of the top management. This is an important element of the organizational framework.

Once an expenditure proposal is approved, it is the onus of the top management to arrange the money. The following sources of money are available:

1. Internal sources
2. External sources

Internal sources are from the profits of the organization or reserves or retained earnings. The external sources are borrowed funds from financial institutions like the banks, share capital, grants from governments, advances from the customers, and so on. Sometimes, the governments in under developed countries come up with schemes to attract investment. In certain countries, they reimburse the entire payroll expenses in respect of the locals employed in the organizations. Some governments give free land; some waive all the taxes on items purchased from the local suppliers and so on. The organizations can benefit from such schemes. Sometimes customers finance the acquisition of equipment. In software development companies, often, the customer supplies the development platform like the SDK (Software Development Kit) and the RDBMS (Relational Database Management System). When a facility is set up exclusively for a large organization, the larger organization provides a part of the investment with or without equity participation. Customers also pay advances for equipment made to their order.

Banks and financial institutions carry out financing as their primary business. They lend money both for working capital as well as for fixed investment. We need to pay compound interest on the capital borrowed from these institutions. These institutions would demand disclosures about the business to ensure that the organization is running profitably and can repay the borrowed money.

When it comes to the share capital, the investors may bring in the entire share capital or raise some of the share capital from the general public. Of course, the investors may co-opt institutional investors who could be commercial banks, private individuals, or merchant bankers, or even the governments. Money is invested in share capital with the aim of getting back money in the form of dividends paid out from the profits. In addition to that, the shares are traded on stock exchanges. So, the investors may also profit from selling their shares when the price appreciates. Raising share capital from general public is strictly regulated by governments and costs quite a bit of money to raise and comply with the regulations.

How to acquire the required funds is a larger question and we do not attempt to answer it here. But it is the onus of the top management to set a system of evaluation and approval of fund requirements, and a system to acquire those funds especially when the internal sources are insufficient to meet the requirements.

2.4 Support Systems

Technical systems are those that are directly linked with delivering a product or service to the customers. For them to work unhindered, support services are necessary. They are:

1. Finance and accounting
2. HR
3. Marketing and selling
4. Facilities maintenance
5. Procurement and warehousing

Finance and accounting functions take care of the finances of the organization. Finance portion of the responsibility is concerned with arranging the required funds for the operations of the organization. They take into consideration all the expected inflows and outflows of money and prepare a funds flow statement periodically and update it regularly. Then they ensure that required funds for the proposed expenditure are made available at the right time.

Accounting function keeps all the books of accounts of the company keeping track of all the inflows and outflows and prepares the end of the year financial statements including the balance sheet, profit and loss statement and so on. Accounting function fulfills the statutory requirements of the organization. All payments are routed through accounting function. Similarly, all receipts are handed over to the accounting function. Accounting function is the custodian of organizational fund and its proper accounting. There are regulations and rules on how to account for the operations of the organization and the accounting function fulfills all those obligations on the part of the organization. When organizational fraud is detected, it is the CFO (Chief Financial Officer) that is responsible in the eyes of law besides the CEO (Chief Executive Officer).

The top management has the onus of setting up a robust finance and accounting function. Smaller organizations are outsourcing the accounting function to organizations specialized in accounting. Finance function is not amenable to outsourcing except in the case of very small organizations. So, it is retained in-house. In medium to large organizations, a robust accounting department is more or less essential and the top management has to institute one in the organization.

For medium to large organizations, which are professionally managed and for those organizations that raised money from the public in the form of share capital, it is essential to have an accredited external auditor. The role of the external auditor is to certify that the organization has diligently kept the books of accounts and that they are accurate. This certificate is used to satisfy the shareholders and the regulatory authorities. The external auditor first checks for internal controls which include an internal auditor. So, in addition to finance function and the accounting function, an internal audit function is also essential for the organization and it is the responsibility of the top management to institute this function and staff it with qualified individuals. These functions are cost centers and top management needs to provide necessary funds to run these functions efficiently.

Finance and accounting functions work directly under the supervision and guidance of the top management. Therefore, the top management not only has the onus of setting up these functions robustly but also to supervise and guide their functioning.

Human resources function is concerned with the employees of the organization. They are the agency that ensures the organizational obligations towards its employees are met. Specifically, HR function performs these functions:

1. Take ownership of all organizational human resources from their induction to separation in all aspects concerning human resources.
2. Be the central agency to receive and resolve all the employee grievances.

3. Devise a career progress path within the organization and administer it.
4. Prepare a human resource plan for the organization projecting the requirement of human resources, the possible attrition and the actions to fill the vacancies.
5. Recruitment of people to fill the positions required to run the operations of the organization.
6. Devise and maintain an efficient and effective employee compensation system.
7. Administer the payroll for the organization and arrange payment to all employees on time in coordination with the accounting department.
8. Devise and maintain policies and procedures to ensure employee retention in the organization.
9. Interface with the employee unions in organizations that have those unions and represent the management in all negotiations besides ensuring that cordial relations exist between the management and the unions.
10. Development of the organizational human resources through structured training programs.
11. Maintain a skill database of the organizational human resources.
12. Devise and institute a mechanism for measuring and maintaining the morale of the human resources.
13. Devise schemes and administer them to maintain and increase the motivation levels of the employees.
14. Devise, maintain and administer an employee recognition, reward and disciplinary system.
15. Undertake the responsibility to ensure that all statutory regulations in respect of organizational human resources are complied with.

There could be other organization specific functions placed on the HR department. HR performs all these functions under the guidance of the top management. So, the top management has to shoulder the responsibility to both institute a robust HR function as well as to guide and supervise it.

Marketing and selling are vital functions of the organization. It is through these two functions, money comes into the organization. Marketing function is concerned with positioning the product in the market, monitor the market for product changes, competition, customer preferences, building brand loyalty, product/service promotion and so on. These functions ensure that prospective customers are favorably inclined toward the company's products and services. An efficient marketing function makes it easier to sell the products or services. Pricing the product or service is also a responsibility of the marketing function. It handles the customer's grievances and complaints and dovetails the feedback into the organization so that corrective actions can be taken to rectify the present issue and preventive actions to ensure that the complaints do not recur.

All in all, marketing takes ownership of the customer and represents the customer to the other departments of the organization.

Selling function performs the actual transaction of selling the product or service to the customer. It interacts with the customer; it provides the service or the product; it actually performs all the paper work, which may just be issuing a receipt or an elaborate set of papers to comply with statutory regulations; it ensures a pleasant customer experience; it actually receives customer feedback and complaints; and it collects the money from the customer for the product or the services rendered. Sales involve an exchange, exchange of a product or service from the organization and money from the customer.

How the sales takes place is something that the top management has to put in place. If our goods or services are delivered to a customer through a retail store the distribution channels need to be built to convey our products to be displayed in the stores. If we need to have our retail outlets on our own like in the case of banks, petroleum products, courier services, or automobiles, we need to set up retail outlets across the geographical region selected for our operations. If our products are such that the sales take place from our premises as in the case of bulk equipment or ships or air craft, we need to advertise and promote to pull the customer to our premises. Sometimes, we need to go to the place of the customer and sell our product or service as in the case of insurance policies.

Thus, the top management has to put in place the sales infrastructure appropriate for our products or services in the geographical region selected by us and ensure that the customers can find and reach us easily when they wish to purchase from us.

2.5 Technical and Managerial Processes

Every organization has two sets of process. One is the technical process set and the second is the managerial process set. In technical process set, the processes for producing/developing the deliverable in the case of product-oriented organizations and delivering the service in the case of service-oriented organizations are included. They would contain technical details and guidelines for carrying out technical activities. In managerial processes, the details of how to manage the technical processes are included.

Processes contain procedures, standards, guidelines, formats, templates and checklists. Procedures contain step-by-step instructions about performing a specific activity. Standards specify specifications about of the materials, technical aspects of the operations and so on. Standards contain such details as the insulation resistance of the insulation on electric wires, the thickness of paint applied on a surface, the torque to be applied while tightening screws or nuts, the type of fasteners to be used in the design, and even the logo of the company. Usually the organizations adopt the set of standards from a standards body like the ANSI (American National Standards Institute), NEMA (National Electrical Manufacturers Association), DIN (Deutch Institut fur Normung) and so on. Additionally, it may develop its own standards to supplement the standards of the selected standards body.

Guidelines are similar to standards but are not as prescriptive as the standards. They are more of a suggestive nature. They include aspects like the methodology for planning, preparation of a progress report, carrying out variance analysis and so on. Formats and templates assist the individual in capturing the information comprehensively. Information needs to be captured while carrying out activities and having a format or a template handy would ensure that no piece of important information is forgotten to be included in the document. Checklists aid in ensuring that the activity is carried out comprehensively. It contains a list of memory tickling entries which can be either checked off as completed or not completed or it may require a yes/no answer against each entry. Checking off or answering yes/no would enable us to ensure that every important step or activity is performed diligently.

Organizations may develop these processes initially or as early as possible to ensure that activities are performed as they ought to be in the organizations. Some organizations do have a set of process assets maintained by a dedicated department. Some organizations may achieve this through a set of SOPs (Standard Operating Procedures). With the release of ISO 9000 series of standards in 1995 and various CMM (Capability Maturity Model) certifications beginning with the SEI's (Software Engineering Institute's) CMM in 1998 emphasized the importance of process quality. Obtaining a certificate of compliance from those institutions is creating confidence in the minds of the prospective customers and many organizations are adopting these standards and models. So, it would be advantageous for an organization to adopt a structured approach to the organizational processes by setting up a process development and improvement group.

It is the responsibility of the top management to decide the best approach to drive the organization based on a defined and continuously improved process and to implement it. It includes earmarking a group of people to champion the process definition, improvement and rolling it out in the organization; provide necessary resources; provide necessary funding; and provide necessary support to those activities.

2.6 Reward and Discipline System

While reward and discipline system forms part of employee motivation handled under the aegis of the organizational HR department, it deserves special treatment, because it is the top management that needs to initiate and institute the framework and the HR department would administer and maintain the system. A robust reward and discipline system would go a long way in maintaining and improving the level of morale in the organization.

It is often the case that these schemes are implemented shabbily in the organization and they result in demotivation than increasing the morale. The reward system, to be effective, ought to have the following characteristics.

1. The reward must be determined on the basis of objective data and the method of determination must be transparent to all the contenders. Every contender ought to be able to know why (s)he did or did not get the reward. The data must be accessible to all. It is the subjective factors that are used in determining the winner that cause heart burn for the losers.
2. The criteria for the reward ought to be achievable by all the contenders. The criteria should not be set so high that no one can hope to achieve the reward.
3. The reward ought to be periodic and regular. If one contender does not get the reward in one iteration, there must be hope for the individual to try and get it the next time.
4. We should ensure that the same individual does not get the reward contiguously even if that individual really scored first. If a person is scoring high consistently, it is better to promote him/her to the next level than keep him/her at the same level. By doing so, a message that no one else can get the reward and another message that consistent good performance does not result in elevation gets across to the contenders.

There can be other criteria but ensuring that the reward system does not suffer from these drawbacks results in achieving enhanced morale.

Similarly, for the discipline system to be effective ought to be practiced following the hot stove theory of Douglas McGregor. It ought to have the following characteristics to be effective

1. It should be implemented without fear or favor.
2. It should be commensurate with the act of indiscipline.
3. Its administration ought to be as close in time to the commission of the act as possible. It should never be delayed.

When a discipline system is robustly built encompassing the above principles, it would be very effective in maintaining and improving the organizational morale.

Simply put, the individuals that perform well should receive a reward and the erring individuals need to receive a punishment commensurate with the misdemeanor. This principle would encourage the enthusiastic individuals and deter the lethargic individuals in the organization.

The top management owns the responsibility of putting in place a robust reward and disciplinary system and to provide resources and funding as necessary to ensure its proper upkeep.

2.7 Improvement of the Organizational Framework

It is one thing to define and roll out an excellent set of processes for the technical and managerial processes for the organization. Left to itself, anything in this world goes in one direction and that is toward deterioration. In any process that is not updated to be in tune with the times, entropy sets in making it ineffective. Therefore, the process assets of the organization need to be periodically analyzed and improved as necessary.

There is a misconception which is not totally unfounded that improvement means making the processes more foolproof and tighter. Many organizations fall into this trap and tightening the processes to such an extent, it becomes so elaborate that nobody in the organization can read, understand or implement it. Every improvement ought to make the process simpler, easy to read, internalize and implement it. Implementation of a process ought to give relief to the persons implementing it. In order to achieve this objective, we need to periodically analyze the results from the implementation of the process, the feedback received on it, the suggestions received for its improvement and the requests for waiver received and improve our process assets to be implementable.

An organization has many areas both in technical and managerial aspects. To analyze the impact of the processes on all these areas is not simple or a job for anyone. It has to have specialists in process definition and improvement. People who are close to implementing them would be biased one way or the other. Some people play upon only the weaknesses and some people extol only the strengths. To evaluate all feedback and trim all the biased hyperbole from it, carry out variance analysis between expected results and the actual results, plot a trend graph, uncover the trend and to come out with real actionable improvements is a job for the specialist.

Therefore, the top management has the onus of setting up a process improvement group and providing it the required resources and funding to make it function effectively.

2.8 Knowledge Management

Knowledge management is an oft neglected responsibility of the top management. True, the knowledge takes a long time for the organization to acquire before it can be organized. But if we do not initiate steps right in the beginning, it would never get off the ground. Initially, there would be no knowledge generated internally but we can start a knowledge repository with external knowledge which can be augmented with the knowledge culled from the operations of the organization. Knowledge repository is a framework where in knowledge pertaining to all aspects of organization is collected and organized in a structured manner and with built in mechanisms to access and retrieve knowledge by the people who need it. As the organization puts in years of existence, it generates a large amount of knowledge which can be dovetailed in to the organizational knowledge repository. A well-organized knowledge repository is very valuable in preventing the repetition of the same mistakes, guiding the new entrants and producing better products and services. It is also highly valued by the acquiring organizations during the times of mergers and acquisitions.

Knowledge management includes setting up a knowledge repository, setting up a group of people to cull knowledge from the large amount of transactions performed within the organization, sift and include relevant and appropriate knowledge in the knowledge repository, and provide the required funding to carry out the tasks efficiently and effectively.

Top management owns the responsibility to set up a robust knowledge management system in the organization and support it through provision of funds and equipment necessary to enable the organization to benefit from it.

2.9 Human Resources Development

Human resources development (HRD) is listed as one of the functions of the HR department. But it deserves separate treatment because it is often neglected and is treated as an appendix of the HR department and relegated to a corner. There is one perception that human resources development is just training conducted internally, with in-house faculty or external faculty and sponsoring employees to public courses and seminars. True, these are indeed the functions and perhaps, important functions of the department.

But human resources development goes beyond that! It is building character in the employee! It is concerned with not only the ensuring the necessary skill set with which the employee can shoulder his/her present and future responsibilities effectively but also making him/her an ambassador for the organization. When a person from the companies like IBM, HP, Honda and such others applies for a job, they are valued very highly. Why? They possess the same education and experience as their competitors but there is something they bring along with them and that is the organizational culture from their employers and that is given a high value.

So human resources development goes beyond the training department and it goes into inculcating organizational values like integrity, commitment, quality of service and so on which are very valuable for the organization not only for the efficient operations but also for the good will in the market.

Top management owns the responsibility for ensuring that the organizational human resources not only have the skills expected of them but also have organizational culture internalized in them. This is achieved through the HRD department and through the processes implemented in the organization, structured training to the employees, knowledge management and other initiatives. This is a continuous activity and top management has to carry out some activities directly and some indirectly.

2.10 Conclusion

I have once observed a management consultant come to an organization way back in the 1980s and ask the top management personnel, thus, "What is your effectiveness area? What is your individual contribution to the organization?" He went on to assert, thus, "Please do not claim any credit for any activity that is performed by your juniors".

Not surprisingly, none could effectively project their effectiveness area. Most said that all the results of the people working under him/her are his/hers.

Some said profitability; some said profit; some said smooth operations; some said decision making; some said marketing; and so on. No one said, facilitation for others to work and excel. No one said building and maintaining an organizational framework which facilitates others to excel.

Here is the answer – the main effectiveness area for the top management is to build a robust organizational framework which facilitates others to work and excel; maintain and update it periodically to retain its effectiveness; and continuously improve its effectiveness. This chapter is aimed at providing the elements of the organizational framework and what is expected of the top management.

Henceforth, we will focus on the activities of the middle managers who utilize this framework and produce the desirable results for the organization.

3

Capacity Planning

3.1 Introduction

Capacity planning is a strategic decision. It is carried out at the time of setting up the organization initially. Whenever a possibility for potential expansion arises, this exercise would again be carried out. Another occasion capacity planning is carried out is when funds are available and potential to utilize the additional capacity is foreseen.

Capacity planning is determining the capability to deliver products or services and the rate of delivery to organizational customers. In organizations, work is performed at multiple workstations with each workstation having a different but definite capacity. Work flows from one workstation to another till all operations required to build the product are completed and the product is ready for delivery to the customer. In manufacturing organizations, the work is organized into multiple shops with each shop specializing in a set of related activities. For example, fabrication shop where parts are fabricated; paint shop where parts are painted; plating shop where parts are electroplated; assembly shop where parts are assembled into the product; wiring shop where electrical wiring is carried out on the product; and packing shop where the product is packed for dispatch to customer. A similar combination can be found in other manufacturing organizations depending on the nature of the product being manufactured.

When we come to software development organizations, people find it difficult to visualize different shops as all the activities are carried out by one class of people, namely, the software engineers. Well, it was so until recently but no more. The complexity of software development increased so much that specializations have slowly crept in. Even if the people are designated with the common designation of software engineer, the specializations have come to differ. Consider the following specializations:

1. **Programmers** – these people develop the software programs necessary for the product. Then with the mushrooming of programming languages, the following technology streams are firming up –

   (a) Microsoft technologies
   (b) Open source technologies based on Unix/Linux
   (c) Mainframes
   (d) Midranges

2. **DBA** (Database Administrator) – these people specialize in designing and managing databases. They model and analyze the data and design the database dividing the data into tables with each table comprising a related set of data. They also design the database in such a way that redundancy in data storage is minimized and the disk space is used in an optimal manner. They also develop views, indexes, stored procedures (programs for data manipulation and retrieval in an efficient manner also referred to as triggers, PL/SQL routines, queries, joins and so on) to make data retrieval fast besides reducing the amount of programming required off the programmers. They also assist the programmers in optimizing the data handling routines in the programs.

3. **User Interface (UI) designers** – with the emergence of web-based software in a big way, the necessity to design a visually appealing user interface became a necessity. Gone are the days of green screens with only characters appearing on the screen. Now the screen needs to contain pictures, icons, differently colored text, buttons etc. Not only that they must be aesthetically designed so that the users are enticed to use the application. Because the applications are web based, the screens also must be displayed very quickly when requested by the user necessitating optimization. So, this is a new class of software developers that are needed by the organization.

4. **Application architects** – originally the computer applications were having just one tier with the software and data residing on the same system and the user interface terminals were connected to the same computer. But with development of client-server computing and Internet, the software, the data, and the user interface may reside on different computer systems. Now it is common for the web-based applications to be having 3 or 4 tiered architectures. This has given rise to another class of software professionals, namely, the application architects (also referred to as software architects).

5. **Software designers** – these individuals carry out detailed design of the software units conforming to the software architecture designed by the application architect, the database and the data manipulation routines designed by the DBA, and the user interface designed by the UI designers. They break the software portion of the application into modules and modules into sub-modules and then into programs and document the software design which then will be used by the programmers to develop software.
6. **Business analysts** – these individuals are proficient in functional domains like banking, finance, materials management and so on and at the same time are well versed with the software development activities. These people interact with the customer representatives and develop the user requirements. They document the user requirements which can be used by the application architects, software designers and DBAs to design the application. These individuals also act as the proxy customer for the development team for issue resolution and final testing before offering the application for acceptance testing to customer.
7. **Testers** – web-based applications significantly increased the amount of testing an application has to be put through. There were about 5 (unit, integration, system, volume and user acceptance tests) varieties of tests during the single tier architectures which have now grown to about 35 varieties. Apart from that, some of the tests have to be conducted multiple times. To ease the process of testing, testing tools are developed and are being used during software development. These testing tools need to be programmed much the same way software is programmed but with much less rigor. So, testing has developed into an independent and essential specialty during software development. The testers plan and conduct software testing to uncover all possible bugs lurking inside the software.
8. **Process specialists** – Most organizations in the present day have adopted process driven working. Insistence by prospective clients on certification by agencies such as the ISO (international organization for standardization) for compliance with their 9000 series of standards or with SEI (software engineering institute) for their CMMI (capability maturity model integration) model propelled the move towards process driven working of software development organizations. Process specialists develop various processes, procedures, standards, guidelines, formats, templates and checklists for use by the software development and quality assurance teams in their working. They also assist the technical teams

in implementing and improvement of the organizational process assets. These people also coordinate with the certifying agencies to conduct necessary audits/appraisals to obtain the certificates and maintain them.

The above persons are line departments as they work in the revenue earning activities. In addition to these persons a software development organization needs persons in the following support departments:

1. Finance department to handle finances of the organization.
2. Marketing department for prospecting, project acquisition, customer relationship management, billing, revenue collection and market research and market development.
3. Human Resources department (HRD) – to provide expert support to software developers to ensure that the organizational human resources are motivated as well as to coordinate all policies, processes and procedures related to organizational human resources including their implementation.
4. PMO (project management office) – In organizations that are execut-ing multiple projects concurrently, it is customary to have PMO to coordinating various project execution activities. PMO takes the role of project initiation, project closure, progress monitoring, exception reporting, metrics collection and analysis in the project execution. PMO also acts as the custodian of records of completed projects. PMO also takes the ownership of organizational workforce and allocate personnel to projects during initiation and receive back persons from completed projects. PMO also maintains the skill database of the organizational human resources for use by various departments.
5. Systems Administration department – This department takes ownership of organizational hardware and software assets including networking and Internet. Their responsibilities include procurement of computer hardware, software and networking components as well as maintaining them. They also maintain the organizational web site. They also ensure security of organizational computer assets from intrusion, hacking and virus attacks.
6. Helpdesk – Two classes of help desks are found in software development organizations, namely external helpdesk and internal helpdesk. External helpdesk receives requests for help (over phone or through email) from customers or prospective customers and provide resolution. Internal helpdesk receives requests from organizational human resources and resolves them. Internal helpdesk assists employees of the organization in

various activities including the official aspects such as queries on salary, allowances, travel and stay arrangements as well as personal aspects such as making payments, reservations, fixing/cancelling appointments, finding rental accommodation and so on.

7. Training department – Training department takes ownership of fulfilling the identified training needs. They conduct in-house training programs or identify public programs conducted by professional training organizations. Then sponsor candidates from the organization to the selected training programs so that required skills can be acquired by the organizational human resources.

In large organizations, all these would be independent departments in their own right. In smaller organizations, these would be combined into fewer departments. Here are some possible departmental mergers found in software development organizations:

1. Project execution personnel would be part of a projects department (or technical department or delivery department) or each project could itself be an independent department.
2. Testing department would be merged either with projects or in some cases with the quality assurance department which also takes ownership of organizational process assets. Sometimes, the quality assurance department itself would be titled as SEPG (software engineering process group).
3. It is also common to achieve independent testing using the programmers themselves and totally doing away with the testers in smaller organizations.
4. Training department and internal helpdesk could be made part of HRD. External helpdesk could be part of project execution department or systems administration department.
5. Systems administration department itself could be merged with projects department.
6. Normally finance, marketing and HRD would be independent departments.

3.2 Basics of Capacity Planning

The first step in capacity planning is the assessment of capacity for a facility.

In certain types of industry, capacity expansion for a facility is very difficult. For example, when you build a steel plant, or a fertilizer plant or a chemical plant, ship building, air craft manufacture, the plant has to be designed for specific capacity. We would not be able to produce more than the designed capacity even if we have more orders. It is not possible to expand capacity temporarily by working overtime or in multiple shifts. To add capacity, we may simply have to build another plant which may take a few years at a great cost. In these cases, the capacity is completely inelastic. In this case, capacity has to be planned with long term in view.

In some other types of industry, we may add capacity but with difficulty. Examples are automobile industry, consumer electronics industry, small fabrication units and so on. In these cases, temporary capacity expansion is possible by going in for multiple shift working, or overtime working. In these cases, capacity is elastic to some extent. Since some elasticity is present, capacity can be planned with mid-term in view.

In some cases, it is possible to easily expand capacity either temporarily or permanently at no great extra cost. Short term temporary capacity expansion can be achieved by both overtime and multiple shift working. Office-type of businesses are a great example of this type of capacity including auditing and assurance services, consultancies and software development. Here capacity is very elastic. It can be planned with short term in view.

When temporary expansion of capacity is required the first option to look at is taking stress by the organizational human resources. We work overtime for a short period until the need for temporary capacity expansion is fulfilled and we return to normal working. Another one is taking temporary staff and working in multiple shifts, if it is feasible.

Subcontracting (outsourcing) is another source of temporary capacity expansion. In the recent times, some organizations are utilizing this source of capacity in a permanent manner. Great example of this philosophy is the 70-70-70 formula used successfully by GE. The formula states that 70% of all work needed in the organization should be outsourced and 70% of the work outsourced must be to facilities dedicated to GE work and 70% of work outsourced to dedicated facilities must be to offshore facilities. It worked well for GE both strategically as well as financially.

Ancillaries are one way of outsourcing. Ancillaries are facilities dedicated to executing the work of a large principal organization. It is the onus of the principal to utilize the entire capacity or the extent of capacity for which the principal assured the ancillary. Ancillary is a dedicated facility for a principal. Even if ancillary has more orders for the part not dedicated to the principal, it has to first execute the principal's orders and then take up outside work. Ancillaries are dedicated facilities used for long term capacity needs.

For short-term and temporary capacity expansion, subcontracting is used. The major difference between ancillaries and subcontracting besides dedication is the costs, capability and capacity utilization of ancillary is known beforehand, but we do not know the same of subcontractors. We make use of subcontracting when we foresee a shortfall in capacity to meet the delivery schedules of our orders on hand. We invite quotations for the needed capacity by either privately approaching facilities having the desired technical capability or through a public tender. A public tender can be in the press or on a B2B (business to business) portal or on our own web site.

One more option used by the engineering industry is to buy components off-the-shelf where such components are available. In software development industry, this is not feasible as yet. We are having COTS (commercial off the shelf) software products as tools for use in development but not to integrate into our software as components. True, application servers and rules engines are being available for integration into our software but these need to be planned before we start software development. They cannot be called as components as a bolt or nut can be. Perhaps, this may become feasible in the future.

3.3 Line Balancing

When discussing about capacity planning, we need to understand the technique of line balancing.

What is a line in the context of capacity planning? It is the sequence of activities in the production of a product or service from the first activity to the last activity. We assume that each of the activities in the line are performed at a specialized workstation(s).

Let us take an example of our software organization and software development work. It is depicted pictorially in Figure 3.1.

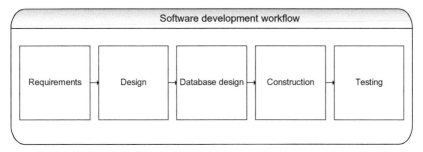

Figure 3.1 Workflow of software development.

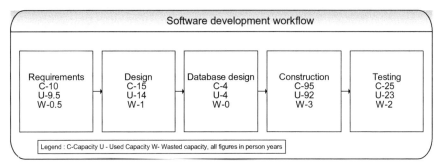

Figure 3.2   Workflow of software development with capacity values.

Now in this diagram, let us add the capacity planned, useful time and wasted time. This is shown pictorially in Figure 3.2.

From these values, we compute two values, namely the line efficiency and the balance delay.

Line efficiency (LE) = total capacity utilized/total capacity available

Balance delay (BD) = total capacity wasted/total capacity available

Total capacity = $10 + 15 + 4 + 95 + 25 = 149$ person-years

Total capacity utilized = $9.5 + 14 + 4 + 92 + 23 = 142.5$ person-years

Wasted Capacity = $0.5 + 1 + 0 + 3 + 2 = 6.5$ person-years

Now LE = $142.5/149 = 95.64\%$

BD = $6.5/149 = 4.36\%$

This tells us that we are having a spare capacity of 4.36% and utilizing 95.64% of the capacity productively.

We need to carry out this exercise for each of the workflows to ensure that the excess capacity is restricted to a minimum.

We have to carry out this exercise between various departments to ensure that no department has unduly excess capacity and no department has unduly inadequate capacity.

Excess capacity drains out monetary resources. The idle people waste the time of productive people besides causing morale to be adversely affected. Therefore, we need to properly balance the capacity between various specialties and departments. Line balancing technique helps us to achieve that objective.

3.4 Utilizing Excess Capacity During Troughs of Workload

Excess capacity results in, mainly due to imperatives of line balancing and some excess capacity becomes inevitable. In addition to that, the lean periods in workload also create temporary excess capacity. We need to utilize the excess capacity in some productive manner so that no one in the organization is idle.

The following are some of the ways for utilizing excess capacity –

1. Preparation of self-study guides for use by new recruits
2. Process improvement activities
3. Conducting training for the organizational resources
4. Assist PMO in various analysis activities like estimated Vs actual values, variances, organizational baselines and so on
5. Assist marketing in preparation of marketing collaterals, proposals, technical negotiations and so on.
6. Perform peer reviews for other projects.
7. Development of tools that assist in improvement of productivity or quality during software development.
8. Any such other useful activities.

3.5 Capacity Planning in the Software Development Industry

Having understood the basics of capacity planning, we can infer that in software development industry, capacity is elastic and temporary capacity expansion is feasible without much hassle. Therefore, we can plan the capacity keeping the short term in view. Normally one year signifies the short term. So, we need to plan capacity of a software development facility such that we need not add capacity for the coming one year. In software development industry capacity has three components, namely, the seating facility, the hardware, software and human resources. While we can add human resources at relatively short notice, we cannot add seating facility at short notice. Therefore, we need to plan the seating facility for a year and acquire it at the beginning of the year itself. Human resources acquisition takes three to four months with a minimum of two weeks, if we maintain a pipeline in recruitment of human resources. Summarizing the above discussion for software development industry,

1. We can plan for seating facility at the beginning of the year.
2. We can plan for human resources as close to the requirement as possible with a minimum of two weeks.
3. Hardware procurement cycle is down to a week now with web-based procurement systems and software can be procured in hours using the download facility. Therefore, capacity planning of hardware and software can be in one week of their actual requirement.

Now let us look at the philosophies of capacity build-up. The following philosophies are used in the industry. While making the strategic decision of capacity planning, we need to consider two variables, namely, the load variance and fixed costs. When we build a certain amount of delivery capability, the cost of maintaining the capacity remains fixed irrespective of the workload placed on the facility. But the workload depends on the market conditions including the competition and is rather unpredictable. When we bid for projects, we cannot be certain of winning every bid and so, we bid on projects twice or thrice the amount of our delivery capability. Sometimes, we may just win projects overshooting our capacity to deliver. Sometimes, we may not win enough projects to utilize our capacity fully.

Now that we are sure that the workload on the facility can vary, we need to define our philosophy of maintaining workforce. The ideal would be to align our workforce, commensurate with the workload. This is depicted graphically in Figure 3.3.

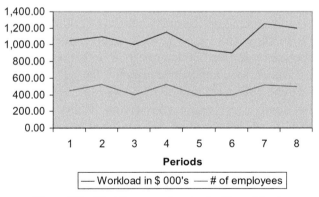

Figure 3.3 Workforce synchronized with workload.

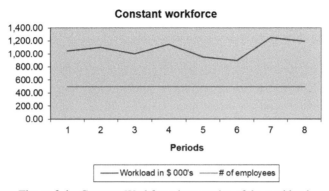

Figure 3.4 Constant Workforce irrespective of the workload.

The second philosophy is to maintain a constant workforce irrespective of the workload on the facility. This is graphically depicted in Figure 3.4

If we adopt the philosophy depicted in Figure 3.4, our fixed costs will remain same during the lean periods of workload and will adversely affect our profits. And during peak loads, there will be stress on the workforce.

It would be better to maintain a workforce that can cater to our lowest workload and expand our capacity temporarily to meet any increase in workload as and when required. This philosophy ensures that our fixed costs are kept low and our profits are maximized. To make effective use of this philosophy, we need to make use of some amount of outsourcing. The outsourcing philosophies are detailed below.

Totally in-house – In this philosophy, we build all the capacity to work on the premises of our organization. No part of the work is allowed to leave the premises of the organization. If we have to expand capacity temporarily, we resort to working overtime or hiring temporary staff from staffing consultants or freelancers from the market.

Partly in-house and partly outsourced – In this philosophy, we build a small in-house staff to carry out the most crucial and sensitive portions of the development and outsource those portions which are not highly sensitive in nature to partners. In this scenario, we do have a dedicated/contracted set of development partners to whom we outsource work regularly at pre-contracted rates. In no case does the 100% work of a project is executed totally in-house.

Need-based outsourcing – In this philosophy, we outsource work on a need-basis. We build a capacity to cater to normal workloads. We try to execute work totally in-house and when the existing capacity is inadequate to meet all delivery schedules, we outsource some of our work which is not of vital importance, to either freelancers or other software development organizations.

Most software development organizations make use of some amount of outsourcing except those few that decide on building up totally in-house capacity. Let us now discuss the process of arriving at the capacity of each department in the following sections.

3.6 Capacity Planning for the Organization

For all other estimates, the order position and capability to procure orders is the starting point. Therefore, let us first learn capacity planning for the marketing department.

The first step is to find out the size of existing market for the type of software we wish to develop. This can be found either through an industry association reports, or economic survey reports or market research consultants or conducting our own market research. Should we consider the geographical limitation while determining the market size? Software development service is such that it can be offered from any global location to any other global location. Once we know the total market size, we can determine the share that we wish to capture from that market. From that size of market, we wish to capture, we can determine the amount of work that we can execute every year beginning from the first year of our operations. Most of this data would be in Dollar amounts. We can convert the Dollar amount to person years of effort by using an average person year (a person year is normally taken as 2000 person-hours) rate. The detailed procedure of how to accomplish this estimation is beyond the scope of this book and you can find more material in the books dedicated to marketing or capacity planning. Let us consider an example.

1. Let us say the total market for developing software using open source technologies was 1 billion Dollars in the previous year with an estimated growth rate of 10% per year.
2. So, the market size would be $1.1 billion for the current year.
3. Let us say that we wish to capture about 1% of this market. That makes our share as $11 million for the first year.
4. Assuming an average person hour rate of $50 (that is $100,000 per person year), this market gives us 110 person-years of work.

Now this figure indicates the number of billable individuals that we need for delivering projects to earn the projected revenue of $11 million per year. This is the first step in capacity planning.

3.7 Capacity Planning for Marketing Department

Now that we have a Dollar amount for the revenue we wish to earn, we can convert this amount to number of projects we need to procure. We can convert the Dollar amount to number of projects assuming the average value of a project. Normally the new organizations are given small projects. Then slowly by establishing an excellent track record, the organization would be able to obtain medium sized projects and grow to large projects. How do we classify, the projects into small, medium or large projects? There are no standards but here are some rules of thumb:

1. A project with an effort of 1 to 6 person-months – normally offered to freelancers. These are mini projects.
2. A project with an effort of 6 to 36 person-months is a small project with an average effort of 18 person-months and a delivery schedule of 3 calendar months.
3. A project with an effort of 36 to 120 person-months is a medium sized project with an average effort of 60 person-months and a delivery schedule of 6 calendar months.
4. A project with an effort of 120 person-months and above is a large project and a delivery schedule of 9 calendar months and up.

Of course, other people may have a totally different set of classifications. Assuming we wish to execute 70% small sized projects and 30% medium sized projects, that will give us a revenue of $7.7 (110 × 0.7) million from small projects and $3.3 (110 × 0.3) million from medium sized projects.

As the value of a typical small project is $150,000 (1.5 person-years @ $100,000) we will need approximately (7.7/0.15) that is about 51 small sized projects.

As the value of a typical medium project is $500,000 (5 person-years @ $100,000) we will need approximately (3.3/0.5) that is about 6.6 or say 7 medium sized projects.

In all, we need to procure 58 or say 60 projects per year.

Now, if we say, our hit ratio (ratio of projects won to bids submitted) is 1 in 3, we need to bid for 180 projects.

Every prospect does not result in asking for a bid. Therefore, until we become a well reputed company, we have to approach more prospects to be able to bid for 180 projects. Assuming that because of our excellent qualification of the prospects, 80% of our prospects ask us for a bid. That makes our prospecting to be 225 (180/0.8) prospects.

Now we have the work load of the marketing department:

1. Prospecting of about 225 opportunities
2. Bidding for 180 opportunities including follow up
3. Customer relationship management for 60 clients
4. Other miscellaneous work

From this workload, we can derive the number of various classes of individuals necessary for the marketing department. Let us assume some norms to make the estimation clear.

1. Assuming that a sales executive would be able to follow 60 prospects (5 prospects per month) per year, we need about 4 sales executives.
2. Assuming that a bidding assistant takes 3 days to prepare a bid, we need to have about 2 bidding assistants (180 bids × 24 hours/2000 hours).
3. We need 2 or 3 people to conduct search for prospects.
4. We need 2 or 3 individuals for promotion.
5. We need administrative support.
6. A head of sales and a head of marketing.
7. Any other people we might wish to add.

Thus, we can estimate the number of people required for the effective functioning of the marketing department. The norms for estimation used in the above exercise are only indicative but not certainly accurate or suggested for use in real life.

3.8 Capacity Planning of Projects (Delivery) Department

To execute projects effectively, we need to plan for the programmers, UI designers, DBAs, application architects, business analysts, software designers, and testers. Now in this class, the essential set is the programmers. Let us take the above data for estimation of capacity here. We already estimated that we need 110 billable resources and 60 (53 small and 7 medium sized projects) projects for execution. Let us assume some data and perform estimation as follows:

1. Assuming an average project team of 6 persons working for 3 months to execute a small project of 18 person months of effort, each team would be able to execute about 4 projects a year.
2. Therefore, we need about 13 teams for 53 projects/4 projects per team) for small project execution.
3. That gives us a requirement of 78 persons (13 teams × 6 persons per team) for small sized projects. These are billable resources.
4. Assuming that a project manager would be able to handle three small project teams concurrently, we need 4 project managers to handle small projects.
5. Assuming an average project team of 10 persons working for 6 months to execute medium sized projects of 60 person-months of effort, each team would be able to execute 2 projects per year.
6. Therefore, we need about 4 teams (7 projects/2 projects per team) for medium sized project execution.
7. That gives us a requirement of 40 persons (4 teams × 10 persons per team) for medium sized projects. These are billable resources.
8. Assuming that a project manager would be able to handle two medium sized project teams concurrently, we need 2 project managers to handle medium sized projects.

Summarizing the above discussion, we need 118 billable resources and 6 project managers. You might have noticed that our original estimate was 110 billable resources and our final estimate is 118! This happens because some capacity gets wasted due to fragmentation of work.

Now, we need to estimate the other skills required to execute the projects. When we do that, the number can rise slightly from the estimated 118.

3.9 Capacity Planning for Software Quality Assurance

Software testing is part of software quality assurance which includes three aspects, namely, the software verification (also referred to as peer review, and software inspection), independent validation (also referred to as software testing) and standards. Verification ensures that the software product is "right" and validation ensures that the "right" product is built. Standards ensure that all development activities would have a minimum standard of quality built-in.

There are various practices for carrying out verification and valida-tion. One practice is to have both the activities performed by programmers themselves, albeit by independent programmers. In most organizations, ver-ification is performed by independent programmers. Very few organizations, if any, have a separate set of persons for carrying our software verification. Some organizations perform all the testing except final testing (which may include tests like system testing, functional testing, load testing, concurrent testing, negative testing etc.) before submitting to the customer for acceptance testing. They would use a separate testing group to carry out the final testing. Some organizations adopt international standards from institutions such as IEEE (institute of electrical and electronic engineers). Others do have an in-house specialist group to define standards for the organization. Those that have an in-house group normally assign this activity to process specialists. The process specialists may be a small group of persons who champion the process inside the organization. But drafting of actual processes and standards would be carried out by the technical people temporarily assigned to the work. The idea behind such a move is that the people performing the work would be best suited to prepare standards and processes. The process group coordinates the activity from designing the process framework, assigning the work of drafting the processes, then performing the quality assurance thereof, pilot implementation, approval and rollout. The process group also coordinates the certification, if desired by the management.

Now how many people are required for these activities?

Let us look at standards and process activity. The difference between the processes and standards required for a large organization would be more elaborate than for a small organization. But the work of these individuals is mainly coordination. Therefore, the team is normally staffed with a min-imum of two and up. During the time when standards are being drafted until rollout of the process, there would be more workload on this section. Therefore, few more people may be assigned on a temporary basis from the projects execution department and then repatriate them back once the work is completed. For regular maintenance and improvement, few specialists are adequate.

For verification activity, the industry thumb rule is 20% – that is it takes 20% of the time that is actually spent on software development. We arrived at 118 people for technical department which necessitates 24 (20% of 118 rounded off to next higher number) persons for verification activity. We need to add this number to technical persons if we follow the practice of having independent programmers performing the verification activity.

When we come to testing, the picture is not so straightforward. There is a thumb rule of 15% for initial testing excluding the final testing but the final testing really depends on the requirements of the individual project. Initial testing includes unit testing and integration testing. In final testing, it is customary to conduct system testing at a minimum, and other tests on a need basis. Such minimum system testing is carried out on one target system. But with web applications, it became necessary to carry out system testing on a minimum of three or four target systems along with load testing and negative testing. If we take the philosophy that initial testing is best performed by the programmers themselves, we need to increase the technical personnel strength by 15%. In our case, it is 18 (15% of 118 rounded to the next higher number) persons need to be added to technical pool.

For final testing, there are no thumb rules. It may take from a minimum of 15% to 50% of the development time! It is also here that testing tools come in handy in reducing the time required for final testing. If we decide to go in for a specialist testing group, we need to go by the number of projects that need to undergo final testing. We provide final testing support for project teams which we estimated to be about 17 in our example. We can assign two people for providing final testing support to each project team. We need to note here that the final testing team would start working only upon completion of a project development work. So, what will these specialists do when the projects are in development? Since these people are testing specialists, these people develop test plans, test cases, testing tool scripts, so that they are ready to start testing as soon as development is completed and finish testing as quickly as possible. Obviously, there will be peaks and troughs of workload for these people due to the case-by-case requirements of projects. Therefore, we may build a minimum team and assign people to this activity from the development team or from outside to cater to peaks and utilize the team for other quality assurance activities during troughs.

Summarizing the above discussion, we have:

1. We add 24 persons additionally to the technical pool to cater to verification activity.
2. We add 18 persons additionally to the technical pool to cater to initial testing activity.
3. We add 34 persons to a specialized final testing department to cater to final testing activity.

The norms used above are only for the purpose of illustration of capacity planning but are not recommended for use. The norms need to be decided based on the specifics of the environment, methods and tools planned for the organization.

3.10 Capacity Planning for PMO

PMO is a support department. It supports project teams in successfully executing the projects. It carries out project initiation, execution monitoring and project closure. During initiation, PMO ensures that the organizational experience in executing similar projects is made available to the project team as well as present norms for productivity and quality. PMO also coordinates with all support departments and ensures that the project has all the SLAs (service level agreements) as necessary from the support departments. During project execution, it monitors project progress and wherever necessary, it highlights issues that need senior management attention and pumps in additional resources required to bring the project back onto track in terms of schedule, productivity and quality. More importantly, PMO ensures that projects have adequate number of human resources at the desired skill level so that project execution can move forward smoothly. In order to perform these activities effectively, PMO maintains the organizational knowledge repository. This repository contains:

1. Records of all projects
2. All software estimates
3. Organizational software and other metrics
4. SLAs of all support departments
5. All proposals for project acquisition
6. Organizational skill database
7. Domain related materials for self study and improvement

How many persons are required to staff an effective PMO?

It takes a senior executive with significant experience in project execution so that the experience would allow that individual to assist project teams effectively. Major time-consuming activities would be project initiation and closure. These are periodic tasks. Project initiation and closure may take between two and four days. Project monitoring and analysis during execution is a recurring task which may take two days in a week. In our example having 60 projects per year, initiation and closure takes a minimum of 240 (2 days for initiation and 2 days for closure for each project) working days and monitoring projects and the recurring tasks takes 104 working days (52 weeks and 2 days per week). We need to add additional person days to cater to miscellaneous tasks for maintaining knowledge repository, internal record keeping, ad hoc support to projects etc. Thus, we have 344 person-days excluding miscellaneous tasks. As we have about 250 (365 less 104 weekend days, yearly paid leave and holidays) working days in a year, it works out to more than one person. So, we may start with a senior person and an assistant and review the workload at the next capacity planning review.

3.11 Capacity Planning for Training

Training department takes ownership for bridging the identified training needs. They need to maintain a database of course outlines, training calendars of public programs conducted by professional training organizations, freelance faculty and in-house faculty for managing the training activity effectively. Their activities include analyzing the projected training needs and determine ways and means of fulfilling those training needs, prepare an yearly training plan for the organization, budgeting and obtaining approvals thereof, finalize the in-house training programs in consultation with technical department and in conformance with the approved training plan, coordinating the training programs, and other miscellaneous activities such as analyzing the training feedback, coordinating with various project managers for nominations for training programs, training needs, coordination with senior management and so on.

The workload really depends of the number of training programs conducted in a year. Some organizations recruit trainees and train them for allocation to projects, and in such organizations, there would be heavier workload for the training department. In other organizations, the workload depends on the number of training programs conducted in-house. Some organizations totally outsource their training programs to a professional training organization. In either case, a coordinator is a minimum requirement. Therefore, we can plan one executive for this department. The capacity for the department can be reviewed after one year considering the organizational philosophy on training and the number of training programs planned for the next year.

3.12 Capacity Planning for Finance

Finance department is one department that exists in perhaps all industries irrespective of their genre and size. Its functions are too well known to need repetition here in great detail. They will have all the finance functions including finance planning, finance management, and accounting. However, since software development doesn't involve input material/parts, there will no purchase and warehouse accounting workload. Whatever purchase accounting that would be necessary is for the purchase of office supplies and such. Payroll and taxes related workload would be there in the same measure as in other industries. An appropriately qualified CFO (chief financial officer) is a certain requirement for a finance department.

Some organizations outsource their payroll work to a professional payroll processing organization. In such cases one executive is necessary to coordinate the payroll activities.

Taxation related work depends on the number of transactions. As raw material purchase is almost non-existent, the only transactions that would attract taxes would be to sales. The rules pertaining to sales and other tax on software differ from state to state and country to country. The workload arising out of compliance to tax rules is also local to the place of business. The capacity required for this role may be arrived at on a case-by-case basis.

Financial planning, budgeting and enforcing budgetary controls is very essential for any organization. Normally this is handled by one or two persons in the software development industry because the workload is low due to the absence of raw materials. With excellent software tools being available for financial management, this responsibility may be handled by the CFO or as a part time activity for a finance executive.

The workload of finalizing accounts periodically has also drastically diminished thanks to financial accounting software tools.

One financial executive with right software tools would be adequate to handle financial planning, budgetary controls and accounts finalization.

It is a mandatory requirement to have an internal audit department for most organizations unless it is a sole proprietary organization. Therefore, we need to have an internal auditor for the organization.

Summarizing the above discussion, we need to have:

1. A CFO
2. One internal auditor
3. One executive to coordinate payroll activities
4. One executive to handle all other functions

We may reevaluate the capacity every year depending on the volume of business and the number of transactions.

3.13 Capacity Planning for Systems Administration Department

Systems administration department maintains the hardware, system software and networking of the organization. They perform the below activities:

1. Coordinate the procurement of hardware, system software and networking components.
2. Install new hardware, system software and networking.
3. Install software development kit on the systems.
4. Maintain all the hardware and software assets of the organizations.
5. Server management of all servers in the organization, including user management.
6. Management of the organizational web site, if hosted in-house.
7. Mail server management including email account management.
8. Backup and restore of organizational data.
9. Support organizational CCB (configuration control board) in maintaining code library (all software code artifacts generated in the organization) and information library (soft copies of project records, metrics, process documents and so on generated in the organization).
10. Normally systems administration department performs the role of configuration controller for the CCB. Configuration controller checks-in the artifacts (both code artifacts and information artifacts) approved by CCB including replacement and ensures that only the current versions of approved artifacts are available to organizational resources.
11. Run an internal help desk for resolving the issues of organizational hardware and software.
12. Maintain the assets register of all the computer assets of the organization, including hardware, software, and licenses. They also maintain the utilization of purchased software licenses in the organizations.
13. Maintain backup media like CDs, DVDs and tapes including issuing them to organizational resources.
14. Ensuring that the organizational systems are protected from viruses, spyware, malware, hacking with appropriate software and hardware guards.
15. Assist the projects during execution with appropriate SLAs for troubleshooting and provision of system resources.
16. In many process driven organizations, the final delivery to customers is effected through the systems administration department. This ensures

that the same set of artifacts is available at the customer as well as in the organizational code library.

Presently, the norm for supporting systems for troubleshooting is one maintenance engineer for every 50 systems. It may appear too much workload for the person but the present hardware is very robust and fails rarely. The maintenance is also very simple in that we replace the components and then get the repairs (if possible) at experts' location, perhaps outside the organization.

Configuration control needs one dedicated resource. Present load is 60 projects a year (5 projects per month). More resources can be added during the next review, if there are a greater number of projects that are concurrently executed in the organization.

We normally dedicate one resource for email management, user management on organizational servers, backups and restore.

We dedicate a person titled Web Master for maintaining organizational web site. This person is necessary whether we host our web site or host it on an outside server. This individual ensures that the web site has up-to-date information; that all links are working; all functions are working and coordinates with developers to make necessary corrections to the web site as and when required. This individual also ensures that the web site is protected from hackers and intruders.

Needless to say, that we need a senior person to supervise all functions entrusted to systems administration department are carried out efficiently and effectively.

Summarizing the discussion, we need:

1. One person to take ownership of systems administration function in the organization.
2. One maintenance engineer for every 50 systems in the organization for troubleshooting function.
3. One configuration controller.
4. One person to maintain email servers and data servers in the organization.
5. One web master for managing the organizational web site.
6. One administrative person to handle the helpdesk and other record keeping of the department.

This would be the minimum requirement, and this can be tailored based on the specific needs of the organization and the functions entrusted to system administration department.

3.14 Capacity Planning for Helpdesk

We have covered internal helpdesk (helpdesk for assisting organizational human resources) in the topic on capacity planning for systems administration department. We need to maintain a helpdesk for assisting our customers when we sell products like Microsoft, or Oracle do. Sometimes we may undertake software maintenance for our customers. In most cases of software development too, we need to provide software maintenance support during warranty period, at a minimum. In all such cases, we need to maintain a helpdesk to support our customers. The helpdesk may be one shift desk or round-the-clock desk. The minimum requirement of resources for a helpdesk is one person for a one-shift desk and three persons for a round-the-clock desk. Beyond that, the workload really depends on the number of calls from customers for support which in turn depends on the robustness of our software and the quality of our user documentation. We need to provide for substitutes when the helpdesk resources take leave. It really becomes cost-effective to outsource external helpdesk to specialist call centers especially when the call center is small. When we provide only warranty support, it is also normal that the person actually providing the support takes all the customer calls for support. In such cases, the need for an external helpdesk is eliminated.

Summarizing, the above discussion,

1. The helpdesk for customer support can be in-house or outsourced.
2. The minimum requirement would be one individual for a single-shift desk and three persons for a three-shift desk.
3. The workload for a helpdesk depends on the number of calls for support which depends on the robustness of our software and the quality of our documentation.
4. If the support provided is only warranty support, then the person providing the support can double up as the helpdesk too.

We can take this as a base and add capacity to meet the workload at the next capacity review.

3.15 Capacity Planning for HRD

Now even though HRD (Human Resources Department) is a support department, it is a crucial department just like finance department. Because all interactions between management and the human resources are routed through HRD, it assumes significant importance in ensuring that the human resources maintain high morale.

HRD takes ownership of human resources at the organizational level. Ideally, HRD should play the role of a concerned parent whose grownup children are out in the world fending for themselves. HRD performs many functions at organizational level to ensure that the human resources are motivated and put in their best performance with great morale as well as ensuring that no project is short of the required personnel.

Planning for human resources is a periodic activity performed once a year and revisited once every quarter to assess the efficacy of the plan. This activity involves collating human resource requirements from all other departments and adding the requirements to fill the vacancies arising out of attrition. Then finalize the plan in consultation with senior management. If projected requirement is in excess of the existing workforce, HRD initiates action to recruit the additional requirement. If the projected requirement is less than the existing workforce, HRD needs to plan for either utilizing the excess workforce in a productive manner or to let them go. This activity is normally handled by the head of HRD with the part time assistance from the HRD staff.

HRD also takes ownership of organizational processes and practices concerning human resources of the organization. This is a one-time activity followed by improvement as necessitated by changing conditions. Again, this is handled by the head of HRD with part time assistance from HRD staff.

Recruitment is a recurring activity in view of the high attrition rate prevailing in the industry. HRD needs to maintain a pipeline to ensure that resources are available at short notice. This activity needs to be carried out in coordination with other departments for scrutiny of resumes and conducting the interviews. One recruitment executive should be able to recruit 50 people per year. We finalize the capacity based on the number of resources to be recruited.

Normally the activities of wage and salary administration, performance appraisals, discipline, career advancement, staff welfare activities, attrition, and other miscellaneous activities into one group and assign personnel based on the workload. Normally one or two executives are assigned for this set for smaller organizations having up to 500 employees. With utilization of computers, the workload is reduced for this area.

Facilities management is often assigned to HRD. The activities included under this head are travel arrangements, housekeeping, security management, office supplies arrangement, guesthouse management, and so on. For a small organization having up to 500 employees two persons would be adequate with one as the minimum.

We need one or two human resource specialists for process definition, process improvement, human resources planning, special assignments such as conducting salary surveys, continuously monitoring the employee satisfaction, hand-holding assistance to other departments, formulating new measures for improving employee tenure in the organization, specialist assistance in handling problem employees and so on.

Summarizing the above discussion, we need to have:

1. One specialist as head of the department for HRD
2. One or two specialists for special assignments
3. One executive for recruitment
4. One or two executives for routine work
5. One or two executives for facilities management

We will be discussing in detail about the HRD activities in the coming chapters.

3.16 Final Words about Capacity Planning

Capacity planning is an involved activity best carried out by industrial engineers. Here, we presented an outline of how capacity planning is carried out to give you an idea about it. Productivity (the rate of achieving results for any work) is the critical factor in deciding the number of employees required for any activity. Productivity norms are unique to the organization based on the methods of working, tool usage, working environment, management philosophies and the amount of automation. Here examples of capacity planning are presented with assumed norms. These norms are not to be used in real life. Then where do you obtain such norms? These norms are available normally with the industry associations, nearest chapter of Industrial Engineering or with the specialists you hired.

One decision needs to be made is – whether we ought to have excess or less capacity. Some organizations like to have capacity a little excess of the requirement. Their philosophy is that "no stress to the employees and no waiting time for the requesters". Some build a little less capacity than is required. Their philosophy is that there is no guarantee that all the projected workload realizes, and human beings are inherently capable of producing more when required and motivated, that would fulfill in case all the projected workload realizes or a little stress is OK for the resources and a little waiting time is OK for the requesters.

You need to define your own philosophy.

4

Human Resources Planning

4.1 Introduction to Planning

A quote attributed to Abraham Lincoln says, thus, *"If I were given six hours to fell a tree, I would spend the first four hours sharpening the axe"*. And another anonymous quote says, thus, *"Nobody plans to fail – they just fail to plan"*. These two quotes highlight the importance of planning succinctly. We cannot state the importance of planning any better than these two quotes.

All works on achieving success in any endeavor begin with the necessity to plan and to plan well. Success may be possible without planning, but planning reduces the risk of failure and increases the chances of success at the same time. Better still, coupled with control (control from the point of view of implementing the plan, including measuring the progress and taking corrective actions), planning brings predictability to the probable outcome of the endeavor besides drastically cutting the risk of failure.

Any human endeavor is prone to failure. The only tool or technique that cuts that risk drastically is *planning*.

The rigor of planning is composed of the following activities –

1. Thinking through the activity – this involves contemplating the activity in terms of what (what are we going to accomplish?), how (the methodology, tools and techniques are we going to utilize), who (what resources do we need?), and when (when exactly are we going to execute each of the tasks?) and then choosing the right alternative for each of these aspects.

2. Recording the plan on paper or a soft copy – this involves capturing all the selected alternatives on paper (or a softcopy) in a structured manner. It helps us to consolidate our thinking and would act as a point of reference for all concerned people when the action begins.

3. Performing quality control activities of verification and validation on the plan artifacts. This involves carrying out the peer review, managerial review and expert review to ensure that our selection of alternatives is appropriate for the proposed endeavor.
4. Organizational support for the planning activity such as, a process, knowledge repository, expert assistance and so on demonstrates management commitment to the planning.

Most managers agree on the importance of planning, but, the question most often asked is, "I sure do plan in my head but is it necessary to capture the plan in a document?" While planning is a necessity, it need not always be put down on paper. For instance, for small short duration endeavors, planning need not be documented. In fact, none of us fail to plan, we do plan. However, it is the degree of rigor (putting it down on paper is an important part of the rigor) which is open for discussion.

Documenting a plan has the following advantages –

1. It can be reviewed by someone besides its author. Alternatively, the planner can review the plan after a little time lapse to see if any important aspect has been missed and thereby improve the plan.
2. We, the human beings, are forgetful especially of small details. Our memory becomes a little hazy after some time. Documenting the plan would ensure that all details are captured for reference by all stakeholders at a future date without having to rely on our fallible memory.
3. A documented plan acts is a point of reference for everyone concerned or involved in the endeavor.
4. It facilitates control of progress and performance evaluation during the implementation.
5. It facilitates validation of the planning parameters by providing a baseline to compare the actual values generated during execution.

Except for very small short duration endeavors, it is a good idea to document the plan. The next question that crops up, once you decide to document your plan, is the level of granularity. The granularity (required detail) of planning depends on –

1. The *duration* of the endeavor
2. The *number of resources* necessary
3. The *complexity* involved
4. The *relationship* between the above three aspects
5. The *geography* of the project

Now consider the sequel to the above statements –

1. Longer the duration, greater is the necessity for increased rigor.
2. If you had all the time in the world to complete the project, the level of rigor and granularity would be minimal. However, in the real world, duration is often constrained increasing the planning rigor.
3. The rigor of planning increases at an increasing rate as the number of resources deployed on a specific endeavor increases. Higher the number of resources employed, the more is the rigor required in planning.
4. Complexity above what is normal for the team increases the need for planning rigor.
5. If the endeavor is executed at one site, it would be less complex in terms of coordination. If the endeavor is to be executed at multiple sites, the necessity for coordination increases. The rigor of planning required is greater for multi-site endeavors than for a single-site endeavor.
6. Different combinations of duration, number of resources employed, geography and complexity, require different levels of rigor in planning.

Before we proceed further, let us also look at what we meant by the "rigor" of planning. The following points define rigor:

1. The time taken for the planning activity. If it is more rigorous we need to spend more time on it.
2. Documenting the plan – Documenting itself indicates that the planning was rigorous.
3. Number of documents – If planning was elaborate as in the case of large scale endeavors involving millions of Dollars, there would be multiple planning documents. These could be, management plan, quality assurance plan, schedule, induction plan, installation plan, deployment plan, commissioning plan, financial plan, exit plan and so on.
4. Quality control of the plan – We can perform, peer review (review by a peer of the planner), managerial review (review by the approver of the plan), and an expert review. If the rigor is more, we may use a group of people in the peer and expert reviews.
5. Approval – multiple levels of approval indicate that the planning process is more rigorous.

The rigor depends on the value of the endeavor that is at stake. If it is a routine monthly planning for a facility in an organization, it would be less rigorous than a one-year plan for the same facility which is less rigorous compared to a five-year plan for the entire organization.

Before we proceed further, let us first define planning.

4.2 Definition of Planning

Planning is defined as *the intelligent estimate of resources required to perform a predefined endeavor successfully at a future date within a defined environment.*

The key terms are

1. **Estimate** – Estimation indicates anticipation using the best guess of the planner. It is likely to vary from the actual values. Estimation indicates that planning precedes performance. Estimation is carried out based on organizational norms (also known as organizational baselines) or the best educated guess when such norms are not available. Estimation is basically a prediction of the future.
2. **Resources** – cover the 5 M's, "Men (people, term "men" is only used for the sake of rhyme), Materials, Methods, Money and Machines (equipment)". Resources are always applied over a period of time (duration). Some resources deplete on utilization like money and materials. Other resources are recurring like people, equipment and methods.
3. **At a future date** – the dates for implementing the endeavor are in the future and are typically decided during the course of planning. Some dates like the ending date are given as inputs to the planning. Other dates have to be determine during the planning exercise.
4. **In a defined environment** – the environment where the work is going to be performed is defined. The environment is either known or is defined during the planning exercise. Any variation in the environment would have an effect on the plan. The Environment refers to a wide variety of conditions including work logistics, workstation design, technical environment, tools, techniques, processes, methods of management, prevailing morale at the workplace, the weather (temperature, humidity, chill which have an impact on the clothing to be used by the people performing the work) and corporate culture to name a few.
5. **Endeavor** – a specific scope of work for which planning is being carried out.

This definition gives us a framework to understand and assimilate the process of planning. The planning process is tailored to suit the specific attributes of the specific endeavor for which we are planning. We gain the following benefits from planning:

1. Planning gives us a peep into the future and all the stakeholders know what is expected of them at a certain point in time. It brings predictability for the people and they can plan their own activities around the plan.

2. It facilitates prediction of the requirement of funds at any point of time. This helps in planning about the procuring the required funds in advance of their requirement.
3. It tells us about the requirement of equipment at any point of time. It tells us how the equipment is loaded with work and if it is overloaded or under-loaded. It aids in better utilization of equipment.
4. It shows how fully people are utilized allowing us to better utilize the available human resources.
5. It gives us the completion times allowing us to make credible commitments to the customer. It can also tell us if we can meet the deadlines set by the customers and thereby improve our plans, and hence, we can meet them.

In some organizations, such as manufacturing organizations, branches of banks, insurance companies, retail stores etc., where the operations are ongoing, schedules based on PERT/CPM would suffice. In organizations using the project model, more rigorous planning involving multiple plan documents would be needed. The amount of planning really depends on the type of organization and the operations involved in delivering the expected results.

4.3 Introduction to HR Planning

HR planning is the planning the quality and quantity of the human resources required by the organization during the planning horizon which is usually one year. It is a periodic activity performed once a year and revisited once every quarter to assess the efficacy of the plan. This activity involves collating human resource requirements from all the departments and adding up the requirements to fill the vacancies arising either due to expansion or out of attrition. Somehow the attrition levels in software development industry are higher than in other industries and are above 10% of the workforce at a minimum. HRD ensures that the requirements projected by other departments are in conformance with the organization. HR planning is carried out at two levels, namely, the organizational level and individual project level.

The basic purpose of HR planning is *to ensure that the business activities are provided with the required human resources, when required, to achieve the business goals of the organization*. This basic objective is broken down into the following subordinate objectives –

1. No activity is left unperformed or underperformed due to non-availability of human resources during the projected plan period.

2. To optimize the cost of human resources so that the organization achieves its profitability goals.
3. Adequate time is available to attract and recruit right talent for the organization.
4. The organization is not burdened with unwanted human resources that may need to be downsized.

4.4 Types of Human Resources

For the purpose of the present topic of HR planning, we categorize the organizational human resources into two categories, namely, the core human resources and the transient human resources.

Core human resources are the pillars of the organization and are employed for longer periods of time. They are experts in the domain in which the organization operates as well as the technical platform in which the organization specialized. The organizations try their best to retain these core human resources as attrition of these core human resources can affect the delivery to customers and affect the very survival of the organization in extreme cases.

Transient human resources are those resources recruited against the requirements of a specific project for finite durations. They may yet be retained in the organization if another project needs human resources with similar skills. Attrition of these transient human resources would not affect the survival of the organization but an untimely attrition of these human resources can affect the timely delivery of the project in which they happen to be working. The organization strives to retain these resources for the entire duration of the project.

HR planning needs to consider both categories of employees and include the need for both in the HR plan.

4.5 Activities of HR Planning

The following are the activities of HR planning –

1. Determine the quantity of core and transient human resources needed for the organization during the projected planning horizon – We take the projected capacity arrived at during the capacity planning discussed in Chapter 2 of this book. The planned capacity to deliver allows us to arrive at the required number of human resources at each of the levels.

We first determine the number of resources required at the working level who actually perform the work and deliver results to the customers. We arrive at this number by dividing the amount of work to be performed by the productivity. Then using the span of control set for the organization to arrive at the supervisory resources. Then depending on the resources so arrived at, we derive the support service personnel required to service the number of human resources arrived at in the earlier computation. Often times, in this planning the support services are overlooked resulting in disparity and impacting the productivity and quality of the output. Then prepare a list of the resources required for the planning horizon. This list will contain the list of skills, the experience needed and the numbers required for each of those categories. A sample format for capturing the information is given in Table 4.1 Each of the departments in the organization prepares this list and passes it on to the HR department who consolidates the requirements and prepares a comprehensive list for the organization which will be utilized during actual planning of the human resources for the organization. Table 4.2 depicts the format for consolidating the requirement of human resources for the organization.

2. Determine the quality (skills and experience) of the human resources needed by the organization during the projected planning horizon – Each of the departments enumerates the quality of the human resources needed for them. They enumerate the quality in terms of the educational qualifications, certifications, experience and any desirable soft skills essential for effective performance of the job. This task needs a thorough knowledge of the work to be performed. The skills have to be described in a general manner so that HR can locate such people. If it is described purely from the standpoint of the organization using organizational jargon, such people may not be available in the market. This will be prepared for each of the positions projected by them. This will help the HR department to locate sparable internal human resources as well as to recruit from outside the organization.

3. Consolidate and arrive at the gap document that gives the details of new human resources needed or the excess of human resources available in the organization – This activity is carried out at the organizational level by the HR department. In this activity they prepare a list of requirements, at organizational level with sub-level of skill set. Now this level is adjusted by reducing the requirements by the following categories of human resources to arrive at the net requirement.

Table 4.1 Format for projecting the required human resources

Department:

Plan Period:

Prepared by: Dt.

Position	Educational Qualifications	Years of Experience	Skill Set Needed	No. of Positions	Present Strength	Variance	Required for Period	Starting Date
<Indicate the organiza-tional position or grade>	*<Indicate the highest qualification needed>*	*<Indicate the minimum years of experience necessary>*	*<Describe the skills in two categories namely the essential and the desirable?>*	*<Indicate how many such people are needed>*			*<Indicate "perpetual" for regular employment or indicate the period in months?>*	*<Give the date from when the employee needs to start working>*

Notes:

1. Show the requirement of regular positions on a separate row and transient positions on a separate row, even if the skill set is the same.
2. In the case of surplus employees, give the date of release in the column captioned as "starting date".

Approved by:

Date:

Table 4.2 Format for consolidating the requirement of human resources

Plan period:			Dt.	
Prepared by:				
Prepared on:				

Requirement of Human Resources by Department				
Position	**Grade**	**Planned strength**	**Existing strength**	**Variance**
Department – 1				
A				
B				
C				
D				
E				
Total				
Department – 2				
A				
B				
C				
D				
E				
Total				
Department – n				
A				
B				
C				
D				
E				
Total				
Organizational total				

Requirement of Human Resources by Position for the Organization

Grade	**Planned strength**	**Existing strength**	**Variance**
A			
B			
C			
D			
E			
Total			

Requirement of Human Resources by Skillset for the Organization

Skillset	**Planned Strength**		**Existing Strength**		**Variance**	
	Regular	**Transient**	**Regular**	**Transient**	**Regular**	**Transient**
A						
B						
C						

(*Continued*)

Table 4.2   (Continued)

Skillset	Planned Strength		Existing Strength		Variance	
	Regular	Transient	Regular	Transient	Regular	Transient
D						
E						
Total						

Surplus employees –
New employees to be recruited from outside –
Approved by:
Approved on:

   (a) Re-deployable human resources declared as surplus by various
       departments.
   (b) Transient human resources that can be moved to regular rolls.
   (c) Promoting employees on regular rolls to next level to shoulder
       higher responsibilities.

4. Determine the strategy to fill the gaps in the requirement and the
   availability of human resources in the organization – This is carried
   out by the HR department in consultation with the senior management of the organization. This would apportion the requirement to
   regular human resources, transient human resources, promotions and
   re-deployment. Numbers are assigned to each of these categories and
   details are worked out.
5. Determine the schedule of the ramping up and ramping down of transient
   resources projected for the planning horizon – Based on the dates of
   requirement given by each of the managers, a schedule of recruitment
   is worked out and dates are assigned for recruitment activities to get the
   human resources well in time to perform the business activities of the
   organization.
6. Determine the recruitment strategies for acquiring the projected
   additional human resources – Recruitment strategies are discussed in
   greater detail in the subsequent Chapter 5 but they include using
   paper/online ads, recruiting agencies, consultants through consultancy
   organizations and any other available alternative. We determine the
   recruitment strategy for each of the requirements and document the
   same.
7. Determine the strategies to ramp down the excess human resources
   available in the organization – After making adjustments, there may still

remain some human resources that may have to be separated. We also determine the strategy to separate them from the organization smoothly. We will discuss this aspect in greater detail in a later Chapter 12. For each category of excess human resources, we determine the strategy to be adopted to separate them from the organization and document the same.

8. Set a budget for the administration of the human resources for the plan duration – Now, we have to set a budget for all the human resources projected in the organization. This will include salaries and perks, bonuses, incentive payouts, pay hikes and so on. We also prepare a funds requirements statement for the plan period so that finance department can make allocation of required funds at the appropriate times. While we may be able to postpone most organizational expenses, employee-related expenditures cannot be postponed without demotivating the employees. Therefore, we need to prepare a credible budget that could be used by the finance department to meet the expenses.

9. Prepare the HR plan for the planning horizon, subject it to quality assurance activities, obtain approvals and release it for implementation – We have to capture all the above decisions in a document. A sample template for HR Plan is shown in Table 4.4. This plan has to be peer reviewed by another manager in the HR department. Then it has to be subjected to managerial review by the head of the HR department. This plan is a very important document whose implementation spreads across the organization. Therefore, it needs to be approved by the CEO (Chief Executive Officer) of the organization.

Of course, this document can span across multiple sheets and may contain some more information appropriate for the organization. The decisions about the untrainable surplus employees and the new employees to be recruited would be recorded in a separate sheet and would go through the normal approval process of the organization but this plan document would form the basis for such decisions.

Once the consolidated requirement of human resources for the next plan period is available, the HR department has to conduct an analysis of the surplus employees to ascertain if they can be retrained and redeployed in the new positions being created in the organization. For those employees, we enumerate the surplus employees in a template depicted in Table 4.3 and arrive at the possibility of retraining the surplus employees.

Table 4.3   Analysis of surplus employee skillset

Employee Name	Qualification (Highest)	Tenure (in Years with the Organization)	Skillset	Matching with New Requirements			Re-Deployability (Yes or No)
				Qualification	Skillset	Trainability	
Total count of re-deployable employees							
Total count of employees not re-deployable							

Management persons, that is the heads of HR, concerned technical departments and senior management persons like the CEO need to consider each case of surplus employees who cannot be retrained for redeployment and arrive at a decision as to what needs to be done in each case. We may have to let some employees go and retrain some even at an extra cost to the organization so that we send a positive message to the rest of the employees that the organization takes care of its employees. But, it is up to the management on how to deal with the surplus staff that has no possibility of redeployment.

Now, once all these details are prepared, analyzed and alternatives frozen, we can finally capture all this in the plan document depicted in Table 4.4 and get it peer reviewed before submission to the management for approval. Generally, HR plans are discussed in a senior level meeting called for the purpose of discussing the HR plan and approving it. All the heads of department would participate in this meeting in addition to those managers who projected sizable new requirements. Usually, the CEO would chair this meeting. This meeting would discuss all the matters in depth and finalize the plan. Once it is approved, the document would be approved and baselined and would be adhered to by all concerned in the organization. Of course, changes can be made to the approved plan if a pressing need arises but only with the specific approvals of the senior management of the organization.

Once this plan is frozen, HR department would begin all necessary actions including planning for the recruitment of new employees, their induction training, retraining for the existing employees and their redeployment, wellness and welfare activities and all such things.

4.6 Process of HR Planning

Every professional organization would have a documented, approved and continuously improved process for performing activities in the organization. It would be so in the case of HR planning too. A process consists of –

1. A set of procedures of how to perform individual activities.
2. A set of formats and templates to capture information uniformly and comprehensively.
3. A set of standards and guidelines to ensure that the activities achieve a minimum level of quality.
4. A set of checklists that can be checked off to ensure comprehensiveness of performance of activities as well as to assist people performing quality assurance activities.

Table 4.4   HR plan for the organization

HR plan for the organization						

Plan period:
Prepared by: **Approved by**:
Prepared on: **Approved on**:

Executive summary
<Give the highlights of the plan including the total strength of the human resources at the beginning as well as at the end of the planning period, the budgeted expenditure, the selected strategies for handling the addition or attrition of resources and so on>

Human resource projection for the planning period:

Skillset	Planned strength		Existing strength		Variance	
	Regular	Transient	Regular	Transient	Regular	Transient
A						
B						
C						
D						
E						
Total						

Human resource strategies
<Here explain the strategies for managing the surplus human resources, if any; acquiring the additional resources required, if any; retraining and redeployment of surplus employees to the extent feasible; tenure of employees including those transient employees; and so on. Also, explain the possibility of retraining and redeploying the surplus human resources attaching the analysis document. >

Financial budget
< The costs and expenses needed for the resources during the plan period needs to be detailed here.>

Training plan
<We shall discuss this aspect in a different Chapter 11 but the yearly training plan needs to be included here.>

Employee wellness and welfare plan
<We shall discuss this aspect in a different Chapter 9 but the yearly plan for these activities needs to be included here.>

Any other relevant organization specific issues
<Here enumerate and explain any other matters that are pertinent to planning of HR activities.>

Usually, the following procedures would be part of HR planning process –

1. Planning initiation procedure
2. Planning procedure
3. Consolidation procedure

4. Plan review procedure
5. Plan approval procedure

Usually, the following formats and templates would form part of HR planning process –

1. Plan initiation format
2. Human resources requirement projection format
3. HR plan consolidation format
4. Minutes of meeting format for capturing the decisions taken during the HR plan review meetings
5. Skillset description format

Usually, the following standards and guidelines would be in the HR planning process –

1. Organizational positions and designations
2. Minimum set of qualifications for each position
3. Attitudinal guidelines
4. Employment of people with disabilities
5. Timelines for performing various activities

There would be a checklist for each of the activities being performed in HR planning. We will discuss these in greater detail in the subsequent Chapter 15 on HR Process definition, improvement and maintenance.

4.7 Issues in HR Planning

Planning in general itself is not an easy task as it involves making assumptions which may or may not be realized. The attrition of employees is one aspect of HR planning that is completely unpredictable. While attrition due to retirement is completely predictable, attrition due to resignations and untimely death/disablement is completely unpredictable. Second aspect of HR planning that is unpredictable is the unforeseen absence of employees. Absence for a day or two can be easily absorbed but unplanned absences of a week or more is difficult to absorb. Here are the issues in HR Planning –

1. The forecast of the requirement of human resources for future is always uncertain. It depends on the marketing forecast which is again an educated guess. If the marketing forecast is erroneous, HR planning would be erroneous
2. The strength of employees is proportional to the size of the expected projects and depends on the productivity. In software industry,

productivity is poorly understood, and credible productivity figures are not available both within the organization or within the industry. The assumed productivity may not be realized in practice.

3. Attrition of employees is unpredictable especially in respect of resignations and deaths. We use assumed figures in this respect which might be erroneous. If the actual attrition exceeds the planned attrition, we may be short of people and if it falls below the assumed attrition, we will be left with excess resources without any work.

4. The individual departmental heads usually take a factor of safety and project higher numbers than required. If those figures are accepted, we may have excess resources. If we cut those numbers downward, then they are very likely to complain that their projects are tardy because HR did not provide them adequate resources. It is a no-win situation.

5. Sometimes, the finance may not be able to provide the necessary funds to acquire the required number of resources as funds are in short supply. The financial constraint may force us to plan for lesser number of resources.

6. The technical executives generally show reluctance toward HR planning as they have to provide credible numbers based purely on assumptions. They become reticent and may have to be forced to participate. While this is surmountable, it delays the activity.

Since there are issues as enumerated above, many argue that HR planning is not essential and a waste of time. They argue that people are available and can be recruited as and when required and similarly, people can be fired when they become excess. Some organizations just do that. They won't use elaborate planning as detailed in this chapter. This approach has one disadvantage. Without advance information and adequate time, we may be able to recruit people, but to recruit people with great talent, planning and time are needed. To recruit good people with talent, we need time to attract, recruit and allow time to become free and join us. Without planning, we would not have time and have to compromise on the quality of the resources. This is a very important factor that supports HR planning.

4.8 Best Practices in HR Planning

All professionally managed organizations implement HR planning as an important aspect of organizational management. Here are some of the best practices used in the HR planning –

1. The organization has a well-defined and continuously improving process for carrying out the activities of HR planning.
2. The senior management of the organization allocates time and resources for this activity besides enthusiastically participating in the activity besides monitoring the implementation of the plan.
3. Allocating adequate duration for the planning activity so that it can be carried out diligently and in a credible manner.
4. The responsibility for the HR planning and its implementation is entrusted to a senior executive in the HR department with proportionate authority and responsibility.
5. Regular analysis and postmortem of the plan vis-a-vis the actual achievement of its implementation ought to be carried out to learn lessons to improve the planning activity in the subsequent cycles and to use the analysis for process improvement.

These best practices facilitate carrying out the HR planning in a credible manner. A robust HR plan will not only provide right human resources for executing organizational projects, but it also optimizes organizational expenses and leads to higher profits.

4.9 Pitfalls in HR Planning

Organizations usually fall into these pitfalls usually under the pretext of budgetary constraints or urgencies. These two reasons are most common excuses for not doing a good job in any organizational activity. Here are some of the common pitfalls that organizations willy-nilly fall into –

1. Not carrying out the HR planning at all or carrying it out in a cursory manner. Sometimes it is viewed as a document-creation exercise to produce a HR plan document rather than view it as an opportunity to take stock of the human resources management in the organization and building right capacity for delivery. A document-creation exercise will produce a plan document that is not credible and would be kept aside the moment it is approved, and the people go about the job without referring to it. This is an ad-hoc approach and would not lend itself to analysis and learn lessons for organizational improvement in the later cycles.
2. Senior management would not allocate resources for this activity as well as not allocate adequate time for carrying out the activity. Senior management would also not spend adequate time on this activity indicating that it is not important to them. This is detrimental to carrying out a credible HR plan.

3. Not having a well-defined organizational process for carrying out this activity. The details of the activity are left to the discretion and expertise of the person championing the planning. If that person is experienced and knowledgeable in the subject, then the activity may be performed credibly. All the same the outcome becomes person-dependent. The organizational experience would not have a bearing on the plan.

4. Not allowing adequate duration is one common pitfall most organizations fall into frequently. Treating this activity as unimportant and spending more time on revenue earning activities would lead to not allocating adequate time to perform this activity in a credible manner. To carry out any activity diligently needs time and we need to allocate adequate time.

5. Entrusting the responsibility of HR planning to a junior executive or as a part-time activity to a very busy executive is another pitfall that organizations fall into. This leads to performing the activity but in a lackadaisical manner. This does not produce a credible HR plan.

6. Not analyzing the actual performance against the plan to draw inferences from the variances and dovetail the lessons-learned into an organizational knowledge repository for future reference.

These pitfalls render the HR planning to be just a document creation activity that cannot be implemented in the organization which in turn would lead to undermining the competency of the organization.

5

Acquisition of Human Resources

Once the HR plan is finalized and frozen for the year, we need to acquire the shortfall in the requirements of human resources so that the organizational projects are supplied with the adequate number of human resources so that the projects can be executed without any hindrance. Now let us discuss the ways and means of acquiring the required number of human resources on time and of the specified quality.

5.1 Strategies for Human Resource Acquisition

Raising and raiding are two popular strategies for acquiring human resources for the organization. Raising strategy involves recruiting people right out of the educational institutions and then train them for the responsibilities to be shouldered by them. Raiding strategy involves recruiting experienced human resources from other organizations carrying out similar work. No organization can adopt either strategy exclusively. An organization needs some raw resources and some experienced resources for their organization.

The experienced employees would hit the ground running. The advantages of recruiting experienced resources are as follows:

1. They need very little training in doing the work. They can begin productive work almost immediately and begin revenue earning work.
2. At most, they need a little orientation to our methods of working and quality assurance and other systems and procedures. The cost of training them is minimal compared to the cost of training raw resources.
3. They can achieve better quality in their work. Experience reduces the propensity to commit errors in their work. They can do it right the first time, most of the time. This leads to overall better productivity and adds to the profitability of the organization.
4. An advantage of recruiting experienced resources is that they can be used to train raw talent recruited by us. The other alternative is to hire

professional trainers from outside at a much higher cost. The external trainers, besides being costly, would not be around when the trainees need clarifications when they put the training to practical work. Our experienced resources would be there to smoothen the rough edges when the trainees begin working on the revenue earning work.

5. They need little supervision. Experienced resources are capable of independent working. They would not disturb the supervisory staff unless there is a serious issue. So, the span of control in the supervisory staff would be better, that is, a supervisor can handle more number of resources than when trainees need to be handled.

However, there are some disadvantages when we recruit experienced resources. Here they are –

1. They are costlier than the raw resources. They need to be paid salaries on par with those working in similar positions in other organizations similar to ours.

2. They take longer time to relinquish their existing job and join us. We need to wait for them.

3. The cost of recruiting the experienced resources is much higher compared with the cost of recruiting raw talent right out of educational institutions. We may need to pay costs of relocation which may involve paying the transportation cost besides other costs like reimbursing the separation expenses from the existing organization, transfer of loans and so on.

4. They bring in their baggage of bad experiences especially with regard to their supervisory staff and may behave in a defensive manner which is not conducive to higher productivity or congenial working.

Recruiting raw talent right out of the educational institutions and training them to shoulder the responsibilities of our organization has its own advantages and disadvantages. Let us now look at them. The advantages are:

1. Their cost is the lowest. Initially, we can pay only stipend during the training period which could span six months to one year out of which some portion is classroom or similar training and the rest is on the job. We can pay them stipend which is near the minimum wage for classroom training as well as on-the-job training. On the job training gives us almost full productivity at the lowest cost. Thereafter, of course, we need to pay them on par with the industry.

2. Recruitment cost of raw talent is also the lowest since we need not pay them any money toward journey or provide them accommodation

or local transport to attend job interview. We may simply go to their educational institutions and conduct a campus recruitment with our own people with no other extra cash outflow besides their journey.

3. We can get the best talent by taking the class toppers at the lowest cost and provide our organization with the best talent within a short time.
4. Since they bring zero baggage of past experiences, we can mold them to adopt our culture and values. They merge with our work environment and culture with zero resistance and with full willingness.
5. Since it is their first job, they would be highly motivated to learn and prove themselves, we would have the highest productivity in their initial years. This would increase the profitability of our organization.
6. If we handle them well, they would remain with the organization for long tenures assuring our organization of stability which assures higher profitability.

Of course, we do always have concomitant disadvantages as in all other cases while recruiting raw talent. Here, they are –

1. The prerequisite to recruiting raw talent is that we need to have experienced resources in our organization that can train the new recruits. All the experienced resources would not have the talent to train raw talent. It is one thing to do a good job but it is altogether a different thing to have the ability to transfer the skill successfully to others. If we do not have right experienced resources internally, we may have to outsource the training to outside agencies increasing our total cost of recruiting.
2. We cannot recruit raw talent at any time of the year. We need to wait for the end of the academic year to get the recruits on board.
3. We may not be able to get all our positions filled by visiting the local educational institution alone. We may need to visit other schools at far off places. If we wish to get the best talent, we may have to stand in queue in premiere institutions competing with organizations like the IBM and Microsoft which results in getting the lowest merit of the institution. It would be, perhaps, better to recruit the best talent from a second ranked institution than the worst talent from the first ranked institution. All in all, we may not get the talent we desire to recruit from the educational institutions and we may need to compromise with the quality of talent we get.
4. During the initial period, the raw talent would not be able to produce the same quality as experienced resources and we need to spend more effort on quality assurance and rectification.

Therefore, we need to weigh our requirements, advantages and disadvantages and then select the mix of resources to be recruited into the organization. Then, the organizations also would be in the following stages of their life cycle:

1. Initial stage – when the organization is just set up and is going into operations.
2. Stabilized stage – the organization was set up with ongoing operations but is still young
3. Mature stage – these organizations were set up sometimes ago and have been carrying out work and earning profits.
4. Declining stage – these organizations reached their peak and are declining with aging human resources and products with diminishing demand.

The organizations in their initial stage would have been just set up with a minimum set of people. They might have been set up by one entrepreneur or a set of entrepreneurs coming together with a common shared vision. More often than not, they would be proficient in the technical aspects of the organization's deliverables. However, they would still need some skills which they might not possess and need to recruit from outside. Being small organization, they would not have the capacity to recruit fresh talent right out of an educational institution and train them. Therefore, the strategy appropriate for these organizations would be to raid other existing organizations and recruit people with experience and the required skills. However, it is very difficult that we get people that exactly match the skill requirements our organization needs. We need to compromise a bit and recruit people with similar skill set and expect them to adjust to our requirements and environment. We need to orient and handhold them in their initial period in our organization. As they are already experienced, they would get adjusted to our environment pretty quickly.

Organizations in the stabilized stage were established some time ago and have set patterns of working and an established organizational work culture. These organizations also would have adequate number of experienced people who can train the recruited raw talent fresh from schools. These attributes of the stabilized organizations enable recruitment of fresh talent right out of educational institutions and train them. So, the best strategy for these stabilized organizations is to recruit talent right from the educational institutions and train them in the required disciplines. Of course, this does not mean that these organizations should not recruit experienced resources. When a new initiative is planned, the organization may not have suitable resources within the organization. They may have to be acquired from outside. But to the extent feasible, they ought to recruit raw talent fresh from educational institutions.

Organizations at the mature stage have long history and would have adequate talent pool that can absorb raw talent. Even when they propose to launch a new initiative, they would have had an R&D (Research and Development) wing that can ensure that the technical side of the initiative is well handled. So, In our humble opinion, the mature organizations need not recruit experienced resources except for special situations. Of course, when they launch a major expansion, they need to recruit experienced resources raiding other organizations. They also recruit experienced persons when they come across an exceptional individual who could help enrich their organization. These organizations do recruit raw talent from educational institutions and train them in the organization to fill the slots available in the organization. Organizations like the IBM, Coca Cola, Pepsi, Microsoft, and Apple come under this category.

5.2 Transient Vs. Regular

Regular resources are those that would be on our payroll. For them, we could foresee continuing work that could occupy 80% of their working hours at least for the coming three years. It is difficult to project the type of skill set beyond three years because the present trend is to have a major change of technology once every three years. We keep resources on regular roaster if the individuals are retrainable because of their education and learning capabilities. We would also keep certain individuals of specialist skills on our regular roaster even if their time is occupied less than 80% of their working hours, because, if they are not available, our organization may suffer severe losses. For example, maintenance staff usually would not have work on their hand all the time but not having them to attend a breakdown would render many more idle resulting in loss of significant number of hours of revenue earning people. Most of the supporting staff come under this category. Database specialists and graphics specialists also come under this category. They too would not have work to occupy their full time, but they are needed within the organization to do vital work. To sum up our discussion, we keep three categories of resources on our regular roaster, namely, –

1. Those for whom we could foresee continuous workload for the next three years.

2. Those that are needed to prevent catastrophes to our organization. These people have specialized skills whom we cannot afford not to have inside our organization.
3. Then we have people with general skills that are needed in our organization and are retrainable as and when needed to fill the vacant positions.

Transient resources are needed for two specific purposes. The first reason we use transient resources are for temporary expansion of capacity. For example, we have a project that needs 100 resources, but we have only 90 that can be allocated to this project. Then we hire ten transient resources to fill the gap in the requirement and release them once their role in the project is finished. Any organization would have peaks and troughs of work load and transient resources are used to cater to the peak work load. Whenever we have a trough in the work load, transient resources are the first to let go. The second purpose of having transient resources is to take advantage of outside expertise that is unavailable in our organization. For example, domain experts come under this category. If we are doing a project in the domain of say radar surveillance software. We may not have expert resources in radar surveillance as our core competence is in software development and not radar surveillance. So, we hire a radar surveillance specialist for the required duration of the project and release that resource after the project is completed. Another example is for reviews. Suppose, we completed the design of a supply chain management system and we hire a material management specialist from outside to review and validate our design and release that resource after the review is completed in all respects. There could be other organizations specific reasons why transient resources are hired. Generally, we do not include very short term, about a month or so, in our HR plans but we do plan for other transient resource requirements.

The recruitment strategy and rigor of recruitment would differ for these two categories of resources we would have in our organization. Now, having completed the understanding of the strategy and background of the recruitment, let us now discuss the sources from where we can acquire the resources needed for the organization.

5.3 Policies for Recruitment

When recruiting human resources for our organization, we need to put in place a few policies regarding the type of human resources we accept into our organization. Here are a few such policies we need to have:

1. Fast track or normal track – When our strategy is to raise our own human resources, we need to decide if we treat all human resources equally in the matter of career advancement. We need to acknowledge that all people with same or similar educational qualifications would have different capabilities depending on their personality traits. Second aspect to consider is that the people who had their education in premiere educational institutions have higher demand and are likely to resign and go away to our competitors if their career is not advancing faster. It is also a fact that the people from non-premiere educational institutes can and perform better than the people from premiere educational institutes, in some cases. If we do not recognize performance and consider only the educational institute, then it is likely that super performers leave our organization and join our competitor. Therefore, usually the organizations use a blend of performance and educational background while advancing careers of their employees. But, it is imperative that we need to have two tracks for career advancement, namely, the fast track and the normal track. We should develop the norms for people to be assigned to the fast track or normal track. Of course, a fast track employee can be moved to the normal track and a normal track person can be moved to the fast track depending on the performance put in by the concerned employee any time during the individual's tenure with our organization.

2. Educational qualifications – Every position in the organization needs to have a specific amount of education. For each position we create in the organization, we need to determine the educational qualification needed to perform the work assigned to that position. We are liable to make type A and Type B errors here. We may over specify the qualification or under specify the qualification. We have to specify the educational qualifications bearing in mind the immediate duties to be assigned as well as the responsibilities that are likely to be assigned in the near future, say, in two to three years of beginning work. That is, we need to specify the educational qualification in such a way that the individual would not be found wanting to discharge the responsibilities of the next higher position, if the need arises. All technical positions need the relevant technical certification if the educational qualification does not include that subject in its syllabus. Care is needed when specifying graduates Vs. post-graduates, and post-graduates Vs. Ph.Ds. That is the confusion lies when choosing adjacent degrees in the education continuum. If our strategy is to raise our own resources, in our humble

opinion, it is better to over specify than underspecify the educational qualification.

3. Persons with disabilities – In the present day, we cannot discriminate employment opportunities based on the disabilities of the person. But some positions need certain specific abilities. When we really consider, all of us have disabilities. Some are short, some are tall; some are obese some are lean; some are great at oral communication while some are poor communicators; in this manner, we differ from each other. We do consider these aspects while recruiting the individuals into our organization. But, we have to be diligent in the matter to ensure that the person is not rejected just because of the disability even though (s)he is appropriate in all other respects and is the best among all other candidates for the position. In software development industry, most common disabilities are not a hindrance for performance. What is needed is that we need to consider our unique situation and set a policy in the matter of recruiting people with disabilities and put in place what disabilities are acceptable and what are not for our organization. Then we need to implement the policy strictly.

4. Social responsibility – Most societies/countries in the world have some sections that were discriminated against for ages. The concept of human equality, secularism, equality of treatment and opportunity are of recent origin. Now, most countries have policies and rules for providing employment opportunity for such disadvantaged sections. They are not likely to be on par with other candidates. We need to recruit them even when they are a little less capable than others when they are suitable for the job. We need to put in place about the sections that come under this category and how much concession we need to give to this section of people and implement that policy scrupulously.

5. Locality – By locality of recruitment, we mean, the distance from which we like to recruit our resources. If the recruitment is restricted to say, 100 miles from our facility then we may refer our preference of our recruitment as local. We can also say local when we restrict our recruitment to our state. If we extend our recruitment to within the country, then we may refer to our recruitment as national. If we extend our recruitment to other nations, then we may refer our recruitment as global. For some positions, we need not extend our reach beyond local level. If we want specialized resources, we may extend our reach to national level. If we want highly specialized resources, we can go beyond our country and recruit at global level. In certain case, there may be a shortage of people

in our country and the government may allow importing of workforce and we can take advantage of it. There is no point in going beyond local level for jobs that pay minimum wage or close to it. For technical positions, we can go to the national level for recruitment. We may go to global level only when there is a severe shortage or the skills required which are of very high order. For trainee positions, we may restrict our recruitment to national level. For skills in short supply at the national level, or for rare skills we can go global. When we need experienced people, we better choose the national level. When we recruit people at the national or global level, we need to pay relocation and resettlement costs. Each has its own advantages and demerits. So, we need to have a policy that specifies the conditions for recruitment at each of these levels.

6. Age – We have a specified range of age to be employable in organizations. It falls within the minimum age for employability and the maximum age of employability. While the age of retirement is flexible in senior positions, it is inflexible to begin working. We have to determine the age requirement for each position. What is needed, however, is that we need to have a policy setting the ages for all the positions in our organization. Some organizations take into consideration the age along with qualification and experience while deciding the rank in the organization. We need to have this policy in place before we go in for recruitment of employees.

7. Gender – Rules do not permit discrimination based on gender for selecting employees, with other conditions being equal. However, there may be positions that are performed by one gender. For example, receptionists, nurses, air hostesses and teachers are filled with females predominantly than with males. Positions of danger are usually filled by males. For example, foot soldiers are predominantly filled by males. But exceptions do always exist. So, we need to set policy guideline of gender employability of positions in our organization which ought to contain where exception can be implemented.

8. Couples – Most organizations comprise of male and female employees. It is possible that committed couples may be employed in our organizations. They may become committed while being employed in our organization. But is it desirable to employ people who are committed to each other even before being employed by us? We have no specific recommendation to offer. Some organizations allow such employment and some frown on it. Some organizations frown on employing relatives

in general too. Each organization and its situation are unique. What is needed is a policy that is appropriate for our organization. We need to define our preference and implement it during recruitment for our organization.

9. Referrals – Our existing employees are our best ambassadors especially in the matter of recruiting new ones. They know our culture and work environment and they have friends who may be apt for our needs. They can do half the work of assessing the candidate even before the resume is received. So, many organizations encourage employee referrals and even pay a nominal sum of money as an incentive for referring a good candidate. But every organization is unique, and we may or may not encourage employees to refer their friends. We need to define a referral policy and make it known to all the employees. Employee referrals cut down the cycle time of recruitment as well as recruitment costs. But it is our prerogative to use it or not for recruiting talent into our organization.

There could be other organization specific recruitment policies like religion, race, ethnicity and caste which could be specific to certain sensitive positions. We need to determine those policies also. We need to implement all these policies while recruiting talent into our organization. Having discussed about recruitment policies, let us now discuss the sources of talent for recruitment into our organization.

5.4 Sources of Human Resources

There are many sources from where we can recruit human resources for our organization. Significant ones are enumerated here.

Educational institutions are the only source from where we can recruit when our strategy is raising our own human resources. Of course, we can place an advertisement asking people without work experience to apply for the position of the trainees. The advantage with going to educational institutions is that we can carefully select the educational institutions and recruit the required numbers. In selecting the educational institutions, we need to consider these aspects –

1. Tier 1 or top-rated university students like to work in tier 1 companies. The students that like to work in IBM, Apple or Google would not like to work in a second-rate company. To entice them, we need to offer them heavier salary packages and proportionate perks which in the long term are not sustainable. We need to assess our own position among

other competitors and select the university that is appropriate for our company.

2. We need to approach the institution either through their placement officer who looks after campus recruitment or the principal of the specific college we are interested sufficiently in advance and book our slot so that we will be among the first to get the opportunity to recruit the best talent from the institute.

3. When we visit the institute, we need to be ready with our corporate presentation which will highlight –

   (a) The background of our organization to give an idea of our reputation to the candidates. We need to remember that the students like to associate and work in a company of which they can be rightfully proud. We should carefully prepare the presentation including the organizational history, any proud achievements of the company or the owners or senior employees, awards won by the company, prestigious orders executed, patenting opportunities and any other organization specific aspects.

   (b) We need to present details of the salary package, career prospects and advancement opportunities, corporate culture, work environment and other relevant aspects to encourage the students to select our organization over others. These will give the students an idea of what compensation benefits they can expect from the organization.

   (c) We also need to present about the special recognition opportunities including awards and rewards including in-house training, knowledge sharing, sponsorship to professional associations, attending and presenting papers in conferences and so on to give an idea of the knowledge enhancement opportunities by joining our organization.

   (d) We may also include a question and answer session to allow the students to ask any question and obtain clarifications on any aspect of our organization.

   (e) Sometimes, we may be in an unfortunate situation of having recent bad publicity. In such cases, we need to prepare well to reduce the impact of such bad publicity on the prospective students rejecting our organization. We need to remember that excellent companies like Toyota had to suffer the ignominy of bad publicity and be ready to face it successfully in case such a need arises.

4. We also need to have a policy of what type of students to select. Should we select from the top-5 ranked students or the lowest rank that can be selected into our organization. Is it desirable to select the lowest ranked student from a tier-1 university as a top ranked student from a tier-2 university may be better? If we select the top ranked student from a tier 1 university, who is likely to receive multiple offers, we are not sure if he would ultimately join our organization. Should we consider the probability of the student rejecting our offer and offer only to those students who would positively join our organization? These are matters we need to have clarity before we go and visit an educational institution to recruit talent.

5. Usually, the campus recruitment takes place in the last semester before the students graduate. So, we need to wait till the graduation to know how many offers are accepted and how many would join our organization. Should we make exact number of offers and take the risk of some positions not getting filled or make a few extra offers and risk having more resources than expected? We also need to have clarity on this aspect.

Visiting educational institutions and recruiting talent from them would restrict our choice to the institutions we can visit. We will not be able to visit all the universities in the country to select the best talent that is coming out of educational institutions. The alternative is to advertise in an appropriate medium and invite applications. We need not advertise in newspapers or TV for this purpose. We may send emails or letters to the placement officers of the institution and request them to pass on this information to their students. If we advertise publicly incurring some expenditure, we can also attract those students that have just graduated but did not accept any campus recruitment offers for reasons of their own. When we recruit people without work experience, we may need to incur expenditure on advertisements and publicity but it may be similar to the expenditure expended on the cost of visiting the institutions for recruitment. The advantage of recruiting using advertisements and publicity is that we get candidates from all over the country. We can also have a link on our company website giving the position or the date of recruitment of trainees and attract applications. Alternately we can accept applications continuously, all through the year, consolidate all such applications and conduct recruitment tests once or twice a year depending on our requirement.

Now, let us turn our discussion to recruiting experienced resources into our organization. While recruiting experienced people, we need to raid other organizations carrying our similar work as that of our organization. However, their methods of working, work environment and organizational culture would be different. While recruiting from other organizations, we could use a targeted strategy or a general strategy. When we use a targeted strategy, we select the candidate organizations from where we could recruit human resources into our organization. The organizations may be classified, for our purpose, into:

1. **Profitably running large organizations** – people in large profitably running organizations usually would not like to leave those organizations because large organizations give them the sense of security of being among large numbers; the anonymity, yet recognition; stability and a clear career path for them gives them a sense of safety and security, which they would not normally like to jeopardize by leaving such organizations. Being identified with a large organization gives them respect in the society. It also provides them access to financing with favorable terms with banks and other financial institutions to acquire large value assets like housing. Yet, they are easy to leave those organization should they decide because the large organizations hardly give them a counter offer because it upsets the remaining people. However, we need to provide them enough incentive in terms of career progression and financial advantage to get them to leave their present organization. One caution though, most people in large organization would not be able to function as efficiently, in smaller and medium organizations as they are used to process-driven methodical bureaucratic way of work culture which small organizations cannot offer. We need to weigh in this risk if ours happens to be a small organization.

2. **Profitably running medium organizations** – people in medium organizations do not enjoy the anonymity a large organization provides. Medium organizations provide them recognition while providing some security coupled with risk. Medium organization has adequate number of people for consultation when necessary but not enough when trouble hits them. It is very difficult to wrest people from a medium sized organization as the top management gives them importance and counteroffer a better salary package than a large organization can. To entice people from a medium sized profitably running organization, we need to give them a much senior position, that is at least two levels above their current level and a matching salary package needs to be offered. The people from a medium sized organization are best fit for all sizes of organizations, be they large, medium or small sized.

3. **Profitably running small organizations** – is small sized organizations, the people usually have a stake at least at managerial levels. Usually, the owner or the top executive directly supervises the managers. So, it would be very difficult to entice them to separate from their organization. There is a disadvantage too – managers in small organizations would not be familiar with the bureaucracy usually prevalent in medium and large organizations. They are used to functioning with oral approvals and orders. Formal approvals or orders, to them, would appear strange and waste of time and effort. They would be a good fit for small sized organizations but would be a doubtful fit for large sized organizations.

4. **Organizations in transition** – many a time, organizations go through transition when acquisitions and mergers happen. The people in both the acquiring and acquired organizations would be apprehensive. Most are not sure if they can retain their position. It is very easy to entice them to our organization. They would be willing to move at almost the same level and salary package to get away from the uncertainty that accompanies the mergers. If we wait till the people are laid off, we can get them much cheaper but they come with a baggage of being taken advantage of their weakness. We would not advocate that. When you are looking for people, we would advocate that you target these organizations in transition. The people you recruit from those organizations would come at a reasonable cost and will be grateful for the opportunity. They will be best fit for large and medium organizations.

5. **Loss making organizations** – When we said loss making organizations, we should not consider those organizations that have just entered the red area in the profit and loss statement. It could be transient and the organization may make profit the very next year. We need to consider such organizations as profit making organizations. The company needs to be in the red for a contiguous period of three years before we can categorize it as a loss-making organization. People working in these organizations are always apprehensive and are continuing there only because they could not get any other opportunity. Of course, there are exceptions to this perception. Some may exhibit exceptional commitment and may decide to sink or swim with the company. But most of the employees would be willing to seize an opportunity even at the same level and salary package. Sometimes, if they are desperate, they may even agree for a cut in pay and demotion in position too. We can hit upon these organizations for our requirements.

6. **Job exchanges/web sites** – There are job exchanges where employers can post jobs and job seekers can post their resumes. Earlier, these were run by individuals or organizations and now there are web sites offering this service. We can look up the existing resumes and recruit the ones that are appropriate for us or we can post our specific requirements and receive resumes. There will be usually some fee, but it is cost-effective and saves us much time in the recruitment cycle time. We can use these for recruiting experienced resources into our organizations.

7. **Job fairs** – Job fairs are organized by universities or job exchanges or in some cases, by the governments. Sometimes, the organizers announce a job fair and then give wide publicity to students and those that are just out of the college. Except in highly reputed universities all the students in the class will not be selected by the visiting organizations. In some educational institutions, no organization visits for campus recruitment for various reasons like accessibility or the reputation of the institution and so on. Even in the lowly ranked institution, some good candidates would be there. All such candidates will take advantage of the job fairs and we can also take advantage of such occasions to get the required human resources into our organization. This is an alternative to campus recruitments. Some organizations regularly recruit from educational institutions and they get priority in recruitment. If we are a one-off campus recruiter, we may not get the priority and get only the second and third rated candidates. For such organizations, job fairs are a good alternative as everyone gets equal priority. We need to depute our recruiting team to the job fair and evaluate the candidates available and recruit the required number of resources. Usually, the organizers intimate the organizations about the job fair, but we can also approach the organizers once we come to know about it. Job fairs are also advertised or publicized as news items from which we can draw necessary information.

8. **Direct resumes** – In the present era, every organization has a publicly accessible web site on which we can place a link for prospective candidates which can be captioned "careers" or something similar and allow prospective candidates to post their resumes. The web site can be programmed to send an automated but courteous response to please the candidate. Whenever we need a resource, we can scan the posted resumes and arrange for recruitment if a resume suits our requirements. We need to have an internal mechanism to receive and carefully categorize and store the posted resumes.

9. **Advertisements** – Advertisements can be gainfully used to acquire resumes for our openings. The advantage with advertisements is that we can attract candidates who otherwise may not available through other means. Good candidates are in a job and do not evince interest unless they are enticed and advertisements do just that. The disadvantage with advertisements is that its cost proportionately increases as we expand our reach from local to national. Another demerit is that the advertisement lasts one day in print media and one insertion in electronic media. Of course, the subject of advertisement is a very large subject and is beyond the scope of this book. With the advent of internet, its popularity and usage in recruitment is increasing multifold and the advertisements are used in a diminishing manner.

10. **Consultants** – Consultants can give us the needed resources within a fraction of the cost we need to spend otherwise. We need not pay any fees to them until we recruit a candidate from them. The cycle time of recruitment is also very short with consultants. They do some of the work that we need to do it ourselves. For example, they do the preliminary scrutiny of the candidates and present us neatly formatted comprehensive resumes. In some cases, consultants handle the entire spectrum of the recruitment activity including the induction program. Consultants can help us in recruiting both fresh talent as well as expe-rienced talent. Some of the consultants also supply us with transient resources. They keep some of the resources on their payroll and allow us to hire them for short periods. Though, we pay a little more for these resources, we need not worry about firing them as they are repatriated to their parent employers.

11. **Individual freelancers** – Individual freelancers are usually those people who somehow did not reach higher levels in their careers. Some people by their very nature are good at one thing only. They like working on their own rather than supervising others. Organizations, usually, cannot pay a higher rate than a range for a specific skill. When a person cannot rise to supervisory level, these individuals come out and set themselves up as freelancers. They are masters of a craft. Especially in computer programming, we find many such people. They are very valuable and useful, especially for specific short assignments. These people charge higher rates than our employees, but their productivity or pace of work-ing is much better than our regular employees. We need to keep contact information of such freelancers in our vicinity along with their skill set. When we get the resource requirements for such short-term high skill

resources, we can call on these freelancers. They are especially useful for temporary capacity expansion or trouble shooting a serious issue.

12. **Government bodies** – Government bodies do have excellent people, especially, in their research and development environments. Governments can and do invest in developing technologies and products that have absolutely no commercial potential, at least, at that moment. For example, who else would think of sending people to the moon and the mars except the governments? Governments are not answerable to any hawkeyed auditors asking the returns on the investment or any bottom lines to show. If their expenditure is more than the income, as it usually does, they just impose more taxes or increase the tax rates. So, if we are looking for resources with rare skills, government bodies are the places to search and locate. Government employees usually do not like to come out of their cocoons but with right enticements, they do come into private sector. They have to be paid much higher salaries and matching perks plus a nice title and a prestigious position.

5.5 Recruitment Methods

For different positions, we need to use different recruitment methods. First question we need to answer is, do we carry out the recruitment ourselves or entrust it to a consultant? In practice, we use both the methods. Each has its own niche in the recruitment. Let us discuss both these methods here.

Recruitment on our own – In this method, we carry out all the activities of recruitment within our organization using our own employees, on our premises. We acquire resumes, we shortlist the candidates for testing and then for interview, conduct interviews using our own executives, then make offers, conduct induction training and enroll the people. While there is no cash outflow for these recruitment activities, there will be significant disruption to the operations of our organization. This will incur opportunity cost for our organization. Many people think that this method is free, because there is no cash outflow from our pocket but is not very correct to think so. The employees used for this work need to put a hold on the work on their hands which may not only delay their work but also delay the downstream activities. To bring the work back to its groove may take days. It may also put an extra burden on other employees who have to shoulder the responsibilities of the person deputed for the recruitment work. This stresses them out. But when we have some trough in our workload as it happens from time to time and we have employees sparable for this work, we can utilize such opportunities to recruit resources. Sometimes, the resources being recruited are of such specialized skill that we wish to keep recruitment confidential, this method has no alternative. Campus recruitment is generally carried out by internal resources. Universities do not like to allow campus recruitment unless they know which organization is going to recruit their students, especially in the case of premiere educational institutions.

Recruitment by consultants – There are consultancy organizations specializing in recruitment alone or as a specialty among other consultancy areas. They recruit resources at a fee. We utilize consultants when we carry out bulk recruitment of fresh talent from all over the country and recruitment of specialist resources. Usually, we limit the geographical area to our city or state for recruitment of resources who are in the working levels. The resources in these levels are those that receive detailed instructions and perform the assigned work. They build the product. In software development, these are programmers that carry out programming of rudimentary routines. But if we are looking for specialist programmers, we need to widen our net and go to the state level and sometimes to the national level. When we look for the best resources, we need to attract talent from all over the country. In such cases, to reach all the target candidates, we need to advertise all over the country. In the days gone by, people had the habit of reading newspapers and we could advertise in one national newspaper and it reached most of the target audience. But now, the habit of reading newspapers is more or less extinct. We can advertise in trade journals and special journals, but their reach is still very limited. We now have consultancy organizations specializing in recruitment of human resources. These consultants gather information of probable target candidates and maintain a database of such organizations. They match our requirements and then make a search of their database and shortlist suitable organizations that are likely to have candidates that fulfill our requirements and reach the target audience with pinpoint advertisements with the least cost. They minimize the advertising cost and reach most of the audience. The flip side is that they charge a hefty fee for the service. When we compare the total cost, the difference may be marginal between ourselves doing it and the consultants doing it but the consultants can get better response. Then we need to decide on the role to be played by the consultants in the total cycle of recruitment. We can restrict their role to the extent of getting right resumes which include getting all the resumes, scrutinizing them and selecting the appropriate resumes with needed skills and in the required numbers. Usually, we consider 3 to 5 resumes for one vacancy so that we will have at least one candidate remaining to join us after putting them through our recruitment sieve. Sometimes we may, based on our exigencies, assign the functions of conducting a written/programming test and conducting the preliminary interview. This will greatly reduce the time to be spent by our internal resources. We may conduct the final interview and make the offer to the selected candidates. We have seen cases where the consultants were entrusted with the total responsibility of selecting the candidates. In such cases, an organizational representative is included in the committee that conducts the final round of interviews giving that person the power of veto.

Of course, the consultants would not have all the skills necessary to technically assess the skills of the candidates but they can get suitable interviewers from different organizations with necessary skills for technical assessment. As an organization, we expect our service provider to perform the task either with internal or external resources and we perform the quality assurance of the process. We ensure that the individuals assessing the technical skills are doing the job well by associating our resource during the technical assessment process.

The advantage with using recruitment consultants is that our resources need not be disturbed from their revenue-earning work besides being cheaper in the total cost than in using internal recruitment. One more advantage is that we can impose strict schedules. When we use internal resources, the recruitment schedule will be a poor cousin in getting higher priority over revenue earning work.

One big disadvantage with using recruitment consultants is that the quality of resources recruited can be poor. If the consultant cannot get right people for technical assessment, the recruited people cannot be up to expected quality. The consultant obviously chooses expediency over perfection and may rush the recruitment process and lose good candidates who may not adjust to the recruitment schedule.

Thus, we need to weigh the advantages with disadvantages and then, make a decision. Recruitment using internal resources and using recruitment consultants have their own advantages and disadvantages. Usually, organizations use a hybrid model by using the recruitment consultants for some of the activities and carry out the vital activities in-house. That is what we advocate too. We would go out on a limb and say that it is safe to entrust the entire process of recruiting trainees straight from universities on campus to recruitment consultants. These trainees would be put through a rigorous training after which the organization would have the option to relieve those trainees whose performance is below the acceptable level.

Head hunting – Head hunting is looking for specialized resources and is generally used to recruit people for senior managerial positions. Usually we do survey the people working in organizations similar to ours and shortlist a few candidates and entrust the work of contacting them and arranging meetings to a consultant. It is very risky to approach these resources directly as it may backfire. If the person contacted is willing to meet us, it is OK but if the person contacted is not willing to meet us, he may report us to his/her employer who would consider this as a hostile act and report us to the industry association or government or any other forum if one such is available. Or, that organization may spread word that our organization is in trouble and cause trouble for us in the stock market or with the customers. So, we entrust the work of contacting the shortlisted candidates to a consultant who would solicit them in a discreet way and disclose our identity only upon ensuring that the contacted individual is willing to relocate. It sure is costly but we need to spend when we are looking for special talent. Head hunting is usually utilized to recruit senior resources in senior managerial positions or some critical technical resources. Head hunting is not used for the bulk recruitment of resources.

5.6 Recruitment Process

By recruitment process, we refer to the steps implemented as part of recruitment of resources into the organizations. There is a full cycle of recruitment and then portions are implemented for different positions based on the position being filled. But first question that needs to be answered is when does the recruitment process begin and what triggers the recruitment action. As discussed in the previous chapter on human resources planning, the approval of the HR Plan provides the number and the type of resources to be recruited. But allocation of budget for the recruitment is the trigger that initiates the recruitment process.

Steps in Relocation Within the Organization

The first action after the budget is allocated, is to identify the organizational resources identified for transfer to other departments and begin the process of transferring them to the new locations. This involves the following steps:

1. Alert the executive supervising the employee identified for relocation to get the work on hand finished and obtain a release date – This would involve arranging a discussion between the identified employee and the

supervisor about all the actions pending from the employee and then agreeing on a schedule for completing the same. Then, freeze those dates and forward the same to the concerned executive in the HR department to take downstream actions such as training and alerting the payroll and receiving department executives.

2. Alert the receiving executive about the joining date of the identified employee – This involves intimating the receiving department head who would in turn, inform the supervisor under whom the relocated employee would be working. The receiving supervisor in consultation with the departmental head would arrange a workstation and determine the duties to be assigned to the transferred resource on joining the department. The receiving department would then indicate their readiness to receive the transferee, to the HR department.

3. If any retraining is identified for the employee being relocated, then plan for the required training and then conduct that training – Some of the relocating employees can be absorbed straight away but some need either reorientation or retraining or minimal amount of new training to make the transferee to be effective in the new role. If there are sizable number needing the same training, then we can conduct an in-company training program and impart the necessary skills. If the number of candidates is small, then we may have to sponsor them to a public training program, is such a program becomes available within the desired time frame. If not, we need to draw up a course outline and evaluation mechanisms for a self-study program and implement it. This may need to be performed under the supervision of the HR department itself. There are a variety of training programs and we will discuss all aspects of training in one of the subsequent chapters.

4. Alert the payroll of the releasing department to discontinue payroll processing from the release date agreed by the department – This involves issuing an order to the agency processing the payroll for the employee about the change. If the agency processing the payroll for both the releasing department and the receiving department, then all that needs to be done by the agency is to just change the charging from the releasing department to the receiving department form the specified date. In large organizations, the transfers can take place between the divisions, in which case, the agencies could be different. In that case the agency processing the payroll for the transferee would transfer the payroll records to the agency which will be processing the payroll for the receiving department.

5. Alert the payroll of the receiving department to begin payroll processing from the date of release from the transferring department. – This involves giving an order to the agency processing the payroll for the transferee from a specific date. If the same agency processes payroll for the releasing and receiving departments, one order would suffice. But, if they are different, we need to issue two separate orders. Once this order is received, the agency will coordinate with the transferring agency and obtain the records and prepare the necessary master file records and make other preparations as necessary to begin processing the payroll for the transferee from the specified date and ensure that the transferee can begin receiving the pay from the first pay date without any issue.

6. Prepare and issue necessary documents for transfer, relocation, and all other required actions. – The HR department has to coordinate all the activities necessary for the identified employee to be released, then trained if necessary and then be assimilated in the receiving department. All these activities need to be performed without causing any interruption to both the departments and the employee. This would involve preparing a few orders and letters as necessary and then issuing them to the concerned agencies on time so that no agency is put to any sort of inconvenience.

Then there could be organization specific activities like exit interviews to obtain the employee feedback about the department and obtaining any information that can lead to improvement and such other activities. All these would become part of the employee relocation activity from one department to another within the organization.

Now let us discuss the activities of acquiring resources from outside the organization.

5.6.1 Steps in Recruiting Resources from Outside the Organization

The following steps would form part of a full cycle recruitment. All these steps may not be implemented in most organizations. Some organizations may have more than these steps depending on their requirement.

1. **Freezing the position requirements** – In the HR plan document, the position requirements are given in a sketchy manner. Now for every position, detailed requirements including the technical skills and other soft skills are elaborated and frozen. The elaboration is usually carried

out by the people who originated the resource request. The elaboration would be at two levels: one for advertising the position and one for selection purposes. Once this elaboration is completed, we are ready to advertise the position in the appropriate channels to invite resumes.

2. **Advertising the position** – Here, the term: "advertising" is used to make the vacant position known to the target audiences. It is not limited to placing an ad in the print or in the electronic media. Now, there are many low/no cost electronic media like the social networking sites, professional networking sites, email lists and so on. We advertise the position in all the relevant media to give wide publicity to the requirements so that we can get as many resumes as possible.

3. **Receiving the resumes** – In the bygone days, receiving was through mail only. Now, it is mostly to an email id. For each position, we set up an email id and direct all the candidates to send in their resumes to that email id. Sometimes, we may display a form on our website and ask the candidates to fill it. Even in such cases, the contents of the form are sent to this email id. We compile all those resumes, position-wise into different folders and consolidate them.

4. **Scrutinizing and shortlisting the resumes** – After the last date for receipt of resumes expired, executives of the HR department would prepare a list of all the resumes received using the format depicted in Exhibit 5.1, usually in an excel sheet or in corporate HR software. Then they carry out a preliminary scrutiny to see if the resume meets the basic criteria set for the position. They separate out those that do not meet the basic criteria specified for that position and send out regret-emails informing the candidate that the resume is not being considered so that the candidate can look for other avenues. Often times, the resume is on the borderline of acceptance/rejection. Such resumes are usually passed on to the originator of the position. They update the list of resumes received list with the rejection information. All the resumes that passed the preliminary scrutiny of the HR executives would be passed on to the originator for a detailed and technical scrutiny along with the list of resumes for that position. The originator would carry out a detailed scrutiny of each of the resumes received and shortlist the candidates for the next step. The originator updates the list of resumes received with appropriate information including shortlisted or not and the reason for acceptance or rejection. This information would be used by the HR department to answer any queries received from the candidates or their representatives.

5. **Conduct a technical written test/programming test including evaluation and shortlisting the candidates for the aptitude test** – Some organizations make a written test mandatory for all except the top few levels. Written test is divided into two levels one being technical test and the other being an aptitude test. Technical test is not mandatory for all levels in the organizations. Technical test is conducted mostly for new entrants either with experience or without any experience but are being inducted into the entry level technical positions. We may have multiple technical specializations in the organization and the technical test would be specific to each of those technical specialties. In the earlier days, we used to conduct a paper-based test but more and more organizations are using computer-based testing as it is quicker and evaluation of answer scripts is automatic and accurate. But one big disadvantage of computer testing is that it cannot asses the essay type of answers of the candidate. Computer based testing is restricted to objective questions and they are not really appropriate in assessing the candidate's knowledge base. When we conduct a test in programming, I am not sure if a computer can be used to assess the quality of the program written by the candidate. The dictum is that if a program works, its quality is acceptable, but we cannot agree with that totally. The delivery of the expected function is but one parameter of the program quality. Maintainability, flexibility and defect prevention are the other dimensions of program quality. While it is a fact that computer testing has become common place for testing the programming skills in the organizations, we would suggest manual assessment in the case of testing the programming skills. Now a day, some code review tools are available in the market and if one is available for the language we are testing, we can perhaps use it. Then we update the list prepared earlier with the results of this test so that the candidates can be shortlisted for the next step.

6. **Conducting an aptitude test and shortlisting the candidates for the technical interview** – In those organizations which make it mandatory, it is the aptitude testing that is conducted for all positions in the organization. Even when the candidate possesses all the needed technical skills, the individual may not deliver the expected results just because (s)he is a misfit in the organization. We test the skills of analysis, comprehension, reasoning, problem solving, integrity, common sense, intelligence quotient, emotional quotient and other traits as necessary for the position. For each position, we determine the important traits and design the test paper and administer it. The design of an aptitude

test paper is a difficult task and therefore, we usually do not design a new paper every time we conduct an aptitude test. We use a test paper for a few years and then revamp the paper. Aptitude testing falls in the domain of psychology and we take the assistance of psychologists in the design of the test paper. We update the list prepared earlier with the results of this test so that the candidates can be shortlisted for the next step.

7. **Conduct the technical interview** – I do not think that there is any organization that lets people into the organization without an interview. It is usually mandatory to conduct a face-to-face interview while recruiting regular resources on to our rolls. However, when we recruit transient resources, often, a telephonic interview suffices. Interviewing is to see if the person projected in the resume and by the written technical test matches the person present before us. Second, personal communicational skills, presentation, gestures and the overall impression the individual creates need to be assessed to ensure that the candidate as a person can fit in well in the existing organizational culture. Once we conduct the interview, we, using the results of this step, update the list of candidates prepared and updated at every step so that candidates can be shortlisted for the next step.

8. **Conducting the aptitude interview** – Technical persons who conduct the technical interview are not well versed in the testing of psychological aspects of the candidate. That is why, the technical interview is supplemented by an aptitude interview, generally conducted by the HR executives. This interviewing will ensure that the candidate does indeed have the aptitude to be a good resource for the organization. Once a resource is recruited into the organization, we need to make that individual an asset for the organization. If the department that recruited the person can no longer utilize that resource, we need to have the possibility of utilizing the resource in another department albeit with some re-orientation or retraining. Sometimes, we may recruit a person especially at a senior level specifically to disrupt the existing culture in the entire organization or a specific department to ring in change for improvement. In such cases, we do look for those traits in the person that can propel the desired change. Personal interviewing is to confirm the resume and the results of the written test. More often than not, the written test brings out the person but sometimes, the written test may not be able to uncover the undesirable traits. Sometimes, the test has shortfalls and sometimes, the individuals can mask their undesirable

traits or even fake the test. That is the reason to confirm the results of the written test with a personal interview. Now, update the list of candidates with the results of this step. Then we shortlist candidates for the final step of managerial interview in the selection process.

9. **Conducting final managerial interview** – Why at all, do we wish to conduct a managerial interview, when the written test, technical interview and the aptitude interview project the candidate as suitable for the position? When all is said and done, it is the manager and the candidate that live together for the foreseeable period. They must be compatible with each other or at least be able to tolerate each other. Why are we using the word "tolerate"? here? Sometimes, we need to take in a person even with incompatible personality because no other individual with those technical skills is available! Some great personalities, each great in himself, could not tolerate each other and fought each other. President Harry Truman and General Macarthur as well as Thomas Edison and Nicola Tesla readily come to mind to bolster this argument. Managers do generally have a large bandwidth in tolerating conflicting view points but there is a limit to that too. Therefore, we conduct a final interview with the manager who would hold oversight over the candidate in the organization. Once a candidate clears this step, an offer would be made to the candidate. We update the list of candidates with the results of this step. Usually, all the candidates pass through this step except in rare occasions. The shortlisted candidates would be made an offer to join our organization.

10. **Conducting the offer negotiation interview** – Every professional organization would have bands for salaries for different levels in the organization. Each of the employees would be put in one band depending on the level. A band will have a range of pay rates depending on the department in which the employee is working. Sometimes, an organization may have different bands for technical and support staff. So, when a new employee is recruited into the organization, the recruit needs to be placed in one of the pay bands. In each band, the pay rates are fixed at the minimum and the maximum pay levels allowed for an employee in the band. When we offer a candidate employment in our organization, we need to offer the individual a level that is either the same level the person is presently in or a level higher than that. Then the salary has to be certainly higher than the one the candidate is receiving at present. In the case of recruiting fresh out of the college, we can offer the minimum pay: rate of the entry level. But in the case of experienced staff, we need

to offer a higher salary than the one being drawn by that candidate. But the vital question is, how much more? First, we need to select a pay band in which the present salary of the candidate can be accommodated. Then we need to fix the pay rate. We can use these guidelines for fixing the pay rate of the candidate –

(a) If the candidate is coming from a loss-making company or is forced to look out or is looking for a better work environment or is coming from a less reputed organization to a better reputed organization, then the individual would be willing to accept the same pay rate or slightly higher because the person has an incentive to accept our offer. In such cases, we can offer say 5% to 10% higher pay rate than the one presently being drawn by the candidate at present.

(b) If the candidate is coming from a comparable company from a comparable work culture, that person would demand a higher pay rate. In other words, the individual has no incentive to come into our organization and therefore would be needing a higher pay rate to lure the candidate into our organization.

(c) If it happens so that we are actively chasing the candidate who is of special significance to our organization, then, we need to offer the individual whatever needs to be given to get that individual into our organization. We need to make him an offer (s)he can't refuse!

(d) Using these guidelines, we discuss the pay and position with the candidate and negotiate both the aspects and come to an agreement.

This needs to be performed for each of the experienced candidates being recruited into our organization.

11. **Making the offer and getting the acceptance** – Once we decide the tentative pay package for the recruited person, we send out an offer letter by email or a letter. The individual may immediately accept the package or email us asking for clarifications or changes or reject our offer outright. When the recruited person asks for changes, we would first assess if the changes asked would result in elevating the position to a next higher level and if it becomes necessary, we may not be able to accommodate the demand immediately. In such cases, we may drop the offer and extend it to the person next in the line or call him/her up and try to negotiate by offering a hike to match the demand in the next iteration of hikes in the company or we may offer non-financial incentives available for special cases in our company. In this manner, we negotiate in a gentle manner with the selected individual and persuade

him/her to join our company. In token of acceptance, we receive a confirmatory email or a signed offer letter or its scanned copy. Along with signing acceptance, we would normally ask for a tentative joining date so that we can plan other activities connected with inducting a new employee into our organization.

12. **Inducting the employee** – The employee joining the organization is a special day, special for the individual and special for us. It is as if a new family member is joining us. It deserves to be celebrated. The new employee comes with a bit of hesitation wondering whether he/she made the correct decision in joining us. We need to put that hesitation at rest and make that individual feel at home with us. It gives confidence to the person that he/she has come to the right place where he/she is welcomed and is valued. Usually, there would be some joining formalities when a new employee joins the organization. The ceremony of joining ought not to be daunting so much so that the individual should not regret the decision to leave the last employment and join us. We have seen some organizations, in which, the new employee is shoved into a cubicle with a bunch of forms to be filled leaving that person alone to fill all those forms. We have also seen organizations in which a HR executive accompanies the new employee and guides him/her to fill a minimum set of two or three forms including the joining report and assures him/her that the remaining forms can be filled at leisure in the next two or three days. Then the new employee is given a guided tour around the facility and introduced to all the concerned people and then given a temporary workstation from which he/she will be operating until a regular workstation can be allocated. Then the new entrant is handed over to the executive who would be supervising the new entrant. Then the supervisor would take over and makes the new entrant feel comfortable and begins working. Once he becomes comfortable, we issue an appointment order which becomes legally binding.

13. **Conducting the induction training** – Should we give an induction training to every new entrant? Would it not put extra burden on the staff? True, these questions are often asked. The purpose of the induction training is not technical in nature. It is intended to give such information to the employee that makes him familiar with the culture of the organization and work environment. We have seen some organizations leaving this induction training to the supervisor but in our humble opinion, it is not the right way. We need not conduct the induction training on the first day but it would be preferable to do it within a month. We need to plan the joining of new employees within a short duration from each other

so that we can conduct the induction training within a month of the first person in the batch joining our organization. Generally, we include the following topics in the induction training program:

(a) History of the company so that the employee identifies him/her self with the founding fathers and takes pride in working with the company

(b) Facilities available to the employees besides what is indicated in the offer letter. These include various financial and non-financial assistance provided by the company and how to access them

(c) Etiquette expected of the employees

(d) Disciplinary rules and office behavior including dos and don'ts

(e) Various HR related procedures and how to utilize them

(f) Grievance handling and escalation mechanisms for resolving grievances

(g) Career progression in the organization

(h) Any other related topic which needs to be known to all employees

14. **Conducting the technical training** – All new employees do not need technical training. All experienced recruits are expected to know how to perform technical activities. Such people need slight reorientation and that can be imparted on the job by supervisors, senior colleagues and organization's defined processes. But when we recruit people fresh out of the universities, we may need to conduct a short technical training program. In the software development field, the universities do not teach programming guidelines, software engineering standards, quality assurance and the development process etc. and we need to train the recruits internally upon their joining our organization. In some cases, when we transfer people internally from one department to another on a mass scale, we need to conduct technical training so they would be able to perform their duties in the new job. All aspects of training, we will discuss in Chapter 11.

15. **Assigning the employee to the specific department** – When do we hand over the employee to the department in which he/she will be working? It depends on the organizational culture. When trainees join the organization, they are usually handed over to the department on completion of their technical training. When experienced resources join our organization, we hand them over to the technical department as soon as they complete their joining formalities. When we recruit general purpose trainees who could be useful in multiple departments like programming, quality assurance, analysis and so on, we generally assess

their attitude and suitability and then discuss the assignment with the individuals. Then we conduct an "assignment consultation" session with the trainees. Sometimes, we may arrange presentations by the heads of different departments. We may also arrange interviews of trainees by the heads of technical departments and allow them to select the trainees into their departments. We generally give the trainees more than one option so they select the one they desire. It is also not untrue that while we give them options, we gently guide them to select the department we desire to assign him/her! After all, trainees are impressionable and all tend to select only one department based on the moment's glamor attached to the department. But we have the onus of ensuring that all departments get the required human resources. Therefore, we need to guide the trainees based on our assessment of their aptitude and suitability. Then we assign them to the selected departments. Experienced resources are recruited against a specific request for a specific department and they would be handed over to that department soon after they complete their joining formalities.

16. **Hand over the employee to the employee relations wing of the HR department** – Once the recruits are assigned to technical departments, the records of the employee would be handed over to the that wing of our HR department which looks after the employee affairs for their tenure. The wing that looks after the employees on regular rolls is called by different names such as Administration, Establishment, Employee Relations or something like that. We hand over the employee records including the initial resume received, the biodata form filled in by the employee at the time of joining, the test scripts, interview records, the forms filled in by the employee as part of joining formalities, internal training records and so on. Once handed over to the Employee Relations wing, the process of recruitment is completed for that position and the recruitment wing moves on to the next recruitment.

17. **Background checking** – Background checking is a contentious issue. Will it be infringement on the privacy of the individual? Is it legal? Will we be violating some law or the other by doing background check? Some governments allow it but some do not! We need to check if it is legal to conduct background checks on the new hires. In some cases, we need not conduct background checks. These positions are usually at the minimum wage and the loss they can cause to the organization is minimal. But it would be desirable to obtain a self-declaration that the individuals are not users of prohibited substances such as hallucinating

Exhibit 5.1 List of Candidates

Serial number	Name of the candidate	Scrutiny (accepted/rejected)	Technical written test	Aptitude written test	Technical Interview	Managerial Interview	Pay negotiation	Offer	Induction	Induction training	Technical training	Assignment	Handover to department	Handover records to employee relations	Remarks

Note: Record the reason for rejection in the cell under the appropriate column. For example, if the candidate was rejected during the technical interview, record the reason for rejection in the Technical Interview column against the specific employee.

drugs, alcoholism, free from addictions, and such other behaviors not permitted in the work place. But in senior positions, the employee can cause substantial and strategic damage to the organization. Therefore, we need to conduct the background verification. One way to be on the right side of law, we need to obtain a clearance from the employee to conduct background checks. When the individual consents, the governments by and large do not have any objection but even that may be prohibited by some governments. We would go on a limb and say do conduct background checks for senior positions. Define a level after which we would conduct background checks. But before proceeding with the background check, obtain the written consent of the concerned individual. We need to have defined and continuously improved process in the matter of background checks and adhere to it unscrupulously.

5.7 Recruitment Pipeline

In industries which have high turnover of employees and software development industry is one such, it is customary to maintain what is called a Recruitment Pipeline. As we have discussed in the previous sections of this chapter, recruitment of external resources passes through multiple stages each of which consumes significant calendar duration as we need to provide adequate time for the candidates as well as our internal technical resources.

Therefore, we usually treat recruitment as a continuous activity especially for entry level resources with one or two years' experience and up to first level managers. At any given point in time, we will have some resumes under scrutiny, some under written test, some under interview and some under offer negotiation. We may even have people in induction too! Because, every stage of recruitment process has candidates, we call it a pipeline. Because, this recruitment is not against a specific requirement against a tight deadline, we take it easy and allow more time in each of the stages of building the pipeline. Here is how we do it:

1. The first step is to acquire resumes. Since the purpose is to build a pipeline, we would not incur any expense toward this activity. We usually place a link on our web site captioned as "Careers" or something similar. That link will lead to a page where we list the hot skills that are in demand in our organizations and encourage candidates to submit resumes to a given email id. We also publicize the needed skills on bulletin boards and internet groups discussing job opportunities such as the Google Groups. When we publicize our vacancies on such media, we may generally not disclose the identity of our organization. We compile all the received resumes and get them scrutinized.

2. We first subject all the resumes to preliminary scrutiny within our HR department to see if they can be passed on to technical scrutiny. While we scrutinize the resumes immediately or as soon as possible for ensuring preliminary qualification, we do not send them for technical scrutiny immediately. We collect a bunch of resumes so that they need at least half a day for the person to scrutinize and then request a technical person to subject them to technical scrutiny.

3. We do not conduct technical written test immediately upon selection in the technical scrutiny but wait till we have a sufficient number of candidates that justify the expense on the technical written test. Once we have that number, we conduct the technical test unless we have some emergency, in which case, we conduct the technical written test as and when required.

4. Even if we conduct the written test, we do not move to the aptitude test unless there is a requirement or we have adequate numbers to conduct the test.

5. We conduct the interviews only when we have a pressing requirement.

We conduct these steps only when we have some spare time on our hands or when we have some idle resources so that we can utilize such idle times productively. We do not spend revenue-earning time just to build the recruitment pipeline. We build the recruitment pipeline using the slack times that the vagaries of organizational working throw up now and then. The advantage of building a recruitment pipeline is that we can quickly recruit the required resources within a short time of receiving the resource request. If we can build a sufficiently large pipeline, we can save money to be spent on advertising and other avenues to attract resumes.

5.8 Intra-organizational Acquisition

Intra-organizational acquisition of human resources is much less complicated and has much fewer steps than acquiring resources from outside the organization. However, there ought to be resources that are sparable who are suitable for the vacant positions albeit with a minimal re-training. Organizations, especially the large ones, do have many people and all of them would not be fully loaded with work. The departments do not wish to declare an employee as not having any work because that resource would be taken off and a request for another resource in the near future would not be entertained by senior management. So, it is common place not to report an employee as sparable. We are sure that many organizations would have resources with some namesake assignment so that they can be declared as doing valuable work. But in project organizations such as software development organizations would always have sparable resources, especially from the completed projects. Resource requirements in project would not be uniform. Initially, only a few resources are needed to ramp up the project. Once the project is fully ramped up, the project would have all the employees needed to complete the project on time. As the project work activities are completed, the resources can be ramped down in proportion with the project completion. Once the project is nearing completion, the resource requirement would be minimal. The resources that have finished their activities would be released as and when their role is completed in the project. In some cases, when employees put in long tenures, they become bored with the present assignment and request for a change of assignment. Sometimes, for some reasons a good resource and a good supervisor do not see eye-to-eye and the resource may request for a change. It is costly to acquire a new resource to replace an internalized employee. So, organizations do not like to fire and hire frivolously and like to retain the existing employees that are willing to relocate. Thus, we have the following reasons for having sparable resources in the organization –

1. Employees getting released from the project because their role in the project is completed.
2. Employees requesting for a change of assignment for various reasons like they want a change of scenario, the supervisor and the employees may not "see-eye-to-eye", or some such other reason.
3. Resources may be under-employed as the existing department is not able to utilize the employee to the optimum level.

Therefore, we can acquire the needed resources from within the organization. The steps involved in acquiring the resources from within the organization are enumerated here.

1. HR department continuously receives requests for change from the existing position and consolidates them into a list.
2. This list is matched against the resource requests received and match the skills of the existing sparable employees and the skills needed.
3. When there is an exact match of skills even with the need for a little reorientation or re-training, the resumes of the matched employees would be sent to the originators of the requests for their approval.
4. When the approval is received, transfer orders are prepared and the employees are transferred to the new departments for absorption after imparting the required reorientation or re-training.
5. The employee database is updated with the change to reflect the transfer of employees from one department to another and the resource request is closed.

The process slightly differs in the case of resources getting released from projects. There is no set date for all the projects to release the resources all at once. The project resources are released as and when a resource completes his/her work. Therefore, these resources need to be handled as and when they become available and expediently as otherwise, they will be idle and disturb other gainfully employed resources. Usually project-based organizations would have a resource allocation cell in addition to HR departments. The resource allocation cell assists the HR department in handling the resource requests. In fact, the resource allocation cell receives all the resource requests and tries to allocate resources from the existing organizational resource pool. They will pass on resource requests to the HR department to acquire from outside the organization only when internal resources are exhausted. In such organizations, resource allocation cells are the agency that approves recruitment either on to regular rolls or transient rolls. Here are the steps in arranging resources to the projects from the organizational resource pool:

1. The project managers raise the resource requests indicating the skill requirement along with the numbers required, the date by which the resources are needed and the possible date by which they would be released and so on.

2. The resource allocation cell matches the requirement with the resources already released from the projects and are waiting for next allocation as well as those that are likely to be released by the date of requirement and prepare a list of resources that can be allocated to the project request.

3. In some cases, there can be a shortfall in the numbers requested in which case, the following actions are taken:

   (a) Allocate a few resources of higher skill level so that their higher productivity (rate of achieving the results) would compensate for the shortfall in numbers.

   (b) We may acquire a tool to significantly increase the team productivity to compensate for the shortfall in numbers.

   (c) We may hire a platform expert to quickly coach the project team to use shortcuts to significantly increase productivity of the project team and thus compensate the shortfall in numbers.

   (d) If the shortfall is marginal like 1 or 2, the projects would take the stress of extra work for a short time, if that is permissible or possible. Level of effort is the second dimension in the Productivity and little increase in the level of effort put in can compensate for the shortfall in numbers.

   (e) Some of the employees would be put on overtime work and compensated extra for extra working to compensate for the shortfall in numbers.

   (f) If there are any trainees in the organization albeit with a slightly less expertise in the required skills, allocate a few more than requested to compensate for the reduced productivity.

   (g) Hire transient resources from freelancers or consultancy organizations and make up the numbers with a clause that these transient resources would be released first.

   (h) If none of the above alternatives is workable, then hire new resources from outside to bridge the shortfall in numbers.

Once allocations are made, resource allocation cell would obtain a sign-off from the project manager that all the requested resources have been received and the project is ramped up as required. One advantage is that all the resources need not be allocated in one installment and this gives a breather to the resource allocating cell to arrange for the requested numbers and skills.

5.9 Pitfalls in Resource Acquisition

The usual practice is to enumerate the best practices first and the pitfalls next. We are deviating from this usual practice with a specific purpose. Recruitment brings in the most crucial resources needed for organizational success. Once a bad resource comes in, we cannot get rid of that individual before some damage is caused to the organization. The amount of damage increases directly in proportion to the level at which the individual was brought in. One characteristic common to all leading organizations that survived passing of the first-generation leadership is that their human resources are of excellent quality both on the technical front as well as on aptitude front. Again, the vice versa, the common characteristic of the failed organizations is that the quality of the human resources is not up to the mark. So, first let us learn the pitfalls in recruitment and then take a look at the best practices. Here are some common pitfalls.

Wrong emphasis – More emphasis placed on either technical or aptitude testing to the detriment of the other. Both are required and balance needs to be maintained between both. A totally technical person without any concern for other human beings or other technical specialties would cause conflict and friction into the organization. While conflict and friction in the organization is unavoidable, we need to keep it at a healthy level. We need to carefully balance these two traits in the individual first by specifying the desired proportions at the stage of defining the position requirements. A person who does not need to interact with other human beings can have less emphasis on aptitude while a person whose primary job is to interface with other people needs to have more emphasis on the aptitude than on technical skills. We have to carefully determine the proportion and then design the testing which needs to be administered meticulously to have the resources apt for the organization. This is one common pitfall most organizations fall into.

Overemphasis on exact match of skills, especially by the HR department – This is a pitfall that most organizations fall into during the stage of scrutinizing the resumes. There is always a blame game between technical and HR persons scrutinizing the resumes. Technical scrutinizers blame HR scrutinizers that resumes that ought to have been rejected are being passed on to them and to prevent this blame, HR scrutinizers become overzealous and reject resumes that differ even by 10%. HR scrutinizers, therefore, look for 100% match which is rare. Other organizations carry out different work in a different environment for different products. How can there be a 100% match between our requirements and the candidate skills, unless, the resume was fabricated? This is the reason why there are quite a few consultancy companies thriving on writing resumes for candidates. We need to strike a balance and educate our HR resources not to expect a 100% match and technical scrutinizers to not blame HR people for passing on all the resumes without their scrutiny. There are organizations in which all the resumes received are simply passed on to the technical scrutinizers without any scrutiny in the HR department. But, HR people like to have a say in rejecting the candidates and do resent taking away their authority to reject. It is for the leadership to tackle this issue and strike a balance that is in the best interest of getting the right resources into the organization.

A & B errors – These errors are referred to by many names. Basically, type A error is accepting the wrong resources and type B error is rejecting the right resources. If you prefer the other way, it is OK. This also is a common pitfall into which organizations frequently fall. There is no organization that is an exception to this. While the type A error causes damage, the resource can be sent away as soon as the mistake is realized which is usually after a significant damage is done. But the type B error is irretrievable. Once we turn down a good resource, that resource permanently becomes unavailable to our organization. That good resource would find another organization and becomes unavailable. Once rejected, our organizational policies do not permit considering a rejected candidate again for some duration, such as 6 months. After 6 months, the candidate is not likely to be interested in an organization that rejected him/her unless the organization is one of the most coveted organizations renowned for being most employee friendly. By enforcing the recruitment process rigorously and meticulously, these errors can be minimized or eliminated altogether.

Biases – It is not an exaggeration to say that every one of us human beings suffers from some bias or the other. These biases surface especially at the time of shortlisting the resumes for testing and during the interviews. How can we prevent biases from keeping the needed good quality resources away from our organization? First, by selecting the right people who can be objective while carrying out their duties and then developing an organizational culture that fosters multiculturalism. These are not easy but that is what we need to do. Of course, we can counsel the individual when a biased behavior surfaces but it is difficult to notice and then it is not easy to effect an attitudinal change in a person.

Negotiation – Negotiation with the selected candidate is a tricky issue. There is a conflict of interest here, the candidate wants the best and the organizational representative wants to give the least possible increase. The negotiation would be for, one is the rank and the other is the pay. We need to consider case by case. Some people are motivated by a better rank; some people are motivated by a higher pay; some people are motivated by higher responsibility; some people are motivated by better work environment; some people are motivated by the field of work; and the list of motivators would go on. We need to assess the motivating factor that propels the candidate to accept our offer, then maximize it and minimize the other aspects to lure the candidate to accept our offer. Negotiation is a skill that the HR people need to master to be better recruiters.

Unresponsiveness – Some organizations fall into this pitfall. Once we acquired resumes, we become silent to the candidate. We do these things:

1. Not acknowledging resumes received – earlier, mails cost us money. Now, we have zero cost email, form letters, mail merge facilities and bulk mailers to automate the process. We can now acknowledge all the resumes received by us automatically at no cost and we need to do this.
2. Not communicating with the candidate – When the process takes longer than usual, we need to communicate with the candidate. Often, we fall silent leaving the candidate to make some assumptions which are usually unsavory. We need to communicate with the candidate until the candidate is either selected or rejected and also need to answer every communication received from the candidate as soon as possible.
3. Not communicating the result to the candidate – Many times, we avoid unpleasantness and therefore, we do not communicate the result to the rejected candidate. Sometimes we do not communicate the result to the selected candidate too. The rejected candidate may send one or two abusive emails to us but if we fail to communicate the result to the selected candidate expeditiously, that individual may choose another organization without waiting for us resulting in losing a good candidate as well as having to repeat the process with additional expenditure. Organizations do not fail because of lack of revenue but because the costs overshot the revenues.

4. Ill treatment to the candidates on our premises – We all need to remember that the candidate is a guest on our premises as the person came to us at our invitation. We should not treat our calls for interview or test as court summons but as invitations. We are not saying that they need to be treated like VIPs but we can easily extend minimal courtesies such as making their entry and exit smooth and allowing them common facilities and their queries attended to courteously. Many organizations treat the candidates as someone seeking alms and not as a prospective future colleague. We need to remember that we too, not long ago, came as a candidate seeking employment in this organization just the way the present candidates have. We need to treat them the way we wished we were treated back then.

We need to avoid or minimize the pitfalls into which we fall into. We fall into these pits mainly because we are not aware of their existence. Many organizations focus on results than on the process using which they were obtained. If we obtain results by unsavory means, the future of the organization is in peril. We need to consciously put in efforts to avoid the pitfalls and get the best resources our organization needs.

5.10 Best Practices in Resource Acquisition

Best practices are the antidote of the pitfalls. By adhering to best practices, we can avoid falling into the pitfalls commonly encountered in resource acquisition. While pitfalls make us commit mistakes and acquire wrong resources, best practices ensure that we get the right resources for our organization. Here are some of the best practices followed in the industry.

1. We advocate definition of a well thought out process for resource acquisition which is internalized in all the stakeholders in the organization. This process needs to be subjected to continuous improvement periodically that too adhering to a well-defined process for improvement. A well-defined process enables even a new comer perform at the level of an expert and an expert to turn in flawless performance. We will deal with the aspect of process definition and improvement in Chapter 15.
2. We need to spend time and resources to design a good test script for aptitude testing for recruitment. Some organizations use a single test script for all recruitment irrespective of the position but, they do define different passing levels and right answers may also differ from position to position. While aptitude testing is a large subject of the field of

psychology, we recommend generation of a large question bank from which the computer can generate a test script according to a set of parameters supplied to it. We can always add questions to it or delete questions from this bank. This will ensure that the test script does not become stale and stays current. A well-designed test script for aptitude testing directed at the position being tested is a best practice.

3. Interviewing is the stage that shuts out or opens our organizational portals to the candidate. Biases and prejudices play a large part here especially in a tight job market. If we can split this to two levels to recommend the result and a tie breaker in case of disagreement in the two levels. We consider it a best practice. Alternatively, some organizations use an external expert who does not have a stake in the result to ensure elimination of bias and prejudice. We may need to select a method that is appropriate to our unique situation but we need to recognize the play of bias in the selection of candidates and try to eliminate or minimize its influence in recruiting right talent.

4. We need to keep up a continuous line of communication with the candidate right from the time we receive the resume to make that individual feel wanted. We need to begin with acknowledging the receipt of the resume, information about the stage of the resume in the process, the invitations for written tests, interviews, communication of the result, negotiation, offer and joining. This is a best practice.

5. Lastly, we need to ensure that we give right importance to all the stages through which a candidate passes from the scrutiny of the resume to inducting the new employee so that no lop-sided selection process hinders getting the right resources into our organization.

6. External technical experts for technical interviews – It is a good practice. An external expert would plug the gaps in the interview conducted by our resource. All experts need not be experts in evaluating the candidates. We may not be able to locate one in our organization or such persons can be busy in their revenue-earning work. So, having an external expert would alleviate such eventualities and ensure that we conduct the best interview and as a sequel, get the best candidates into our organization.

We need to recognize that recruitment is a very important aspect of getting the right human resources who form the backbone of the organization and give a personality to the organization and build its reputation in the market and the world at large.

6

Managing People at Work

6.1 Introduction

Organizations acquire human resources to perform some work. People come to organizations primarily to earn money to take care of their needs. People spend all their time doing work or performing other work-related activities besides attending to personal needs. When we discuss about people management, managing people at work assumes paramount importance. The term "management" was defined as getting things done and the activities performed under its umbrella were supposed to be planning, organizing, staffing, coordinating or controlling and leading or directing. But with the changed scenario, this definition underwent metamorphosis. Now the work of a manager is not just to get things done but to obtain results. The activities of planning and coordinating are still the primary activities of the manager.

The work scenario has metamorphosed in the recent times especially with the advent and low-cost availability of high-speed internet moved the workstations from offices to the homes. Especially in software development and other information-processing desk jobs, there simply is no reason for people to come to office. They can easily perform their work from the comfort of their own home and have some amount of flexibility to choose their own working hours. This also works to the advantage of the organizations in that the costs of owning and running an office space are simply eliminated or drastically cut down adding to the profitability. Now it is in this scenario, that people have to be managed at work who, physically, may not be present.

6.2 Nature of Work in Software Development Organizations

Software development organizations are basically project-focused organizations. That is, they are not repetitive mass or batch production organizations. When cars or televisions are produced, they produce the same car or television continuously for some duration, say a quarter year. A worker performs the same set of operations every day for that quarter year. While the product is the same, it needs to be built repeatedly to achieve the desired number of products. Therefore, detailed instructions of what needs to be performed and how to perform with what tools etc. needs to be explained once for each batch and the individual carries out the duties daily without bothering the supervisor till the end of the present batch and the next batch with another product begins.

But in software products, it takes time to build the first product and the desired number are produced by an automated process of making copies of the product on CDs or DVDs. So, the individuals work on a product only once at any given point in time. So, there is a need to explain the work in detail every time an activity is assigned to the software developer. Of course, we have similar type of project organization in manufacturing too. They are called job-order production organizations. Basically, production is classified into two classes, that is, made-to-warehouse and made-to-order. Made-to-warehouse organizations are characterized by the fact that the rate of production exceeds the rate of demand for the product in the market. So, they produce a number of products in a batch and stock them in their warehouses from where they fill the orders as they come in. When the stock is exhausted, they run another batch. Made-to-order organizations are characterized by the fact, that there is no standard product to make and stock. For example, let us take an organization like the General Electric which manufactures electricity generating stations. There is no standard generating station. It depends upon the design of the generating station to determine what equipment is needed to build the station. So, they do not begin designing until they receive the specifications from the customer as to what needs to be designed and built. Ship building is also similar. So are manufacturers of fertilizer plants, chemical plants, pharmaceutical plants and so on. There are many manufacturing organizations that follow this type of production system.

Our software development is akin to this type of made-to-order system. In fact, software development is a made-to-order or developed-to-order system if you prefer this taxonomy. We develop a software product once but maintain it over its life time. Software maintenance is carried out on a product that is in production to restore it back to its working condition or to make it work in the changed conditions. As software does not have moving parts, wear and tear do not come into play here but the environment or the configuration of the hardware on which our software functions change due to environmental changes like new viruses, new tools, changed protocols, upgraded platform, upgraded Operating system and so on. These changes can and do affect the functioning of our software and we need to upgrade our software to function efficiently on the changed configuration. One other very important difference in a software product is that a physical product cannot be changed beyond a marginal limit, to perform new functions. A ship that was built to carry 30,000 tons may be extended to carry 30,100 tons using the safety factor built into the design, but it cannot be changed to carry 40,000 tons. A software product, on the other hand can be extended without any limits.

So, software development is carried out using a project model. Let us now see the characteristics of a project.

6.3 Executing a Software Project

A project is developing or building a non-repetitive product. In our case, it is developing a software product. Software product comprises of a number of subassemblies and components. Subassemblies can comprise of sub-subassemblies or components. A component is the smallest element of a product. In software development, a component is a program, a data file or a database table. While we use the term "program", it includes scripts, macros, agents, objects, and so many other proprietary terms used by different organizations and the primary functions performed by those are the same as that of a program. Sub-subassemblies and subassemblies also consist of software components. Project work consists of

1. Building these components
2. Testing the components and removing the defects therein
3. Assembling them into the sub-subassemblies and subassemblies
4. Testing the sub-subassemblies and subassemblies and removing the defects therein
5. Assembling all the sub-subassemblies, subassemblies and other needed components into the product
6. Testing the product and removing the defects uncovered
7. Building the deliverable
8. Conducting the acceptance testing by the customer or the end-user
9. Develop the documentation necessary for efficient usage and troubleshooting of the product during operation
10. Installing the product on the target hardware
11. Training the end-users in the usage of the system

12. Trial runs to give hands-on training to the end-users
13. Cleaning the junk data created during the trial runs
14. Roll out the system to practical use
15. Hand over the product to the software maintenance team.

These activities complete the project of developing a software product. Then the software maintenance of the product is commenced. Software maintenance is like running a restaurant or a post office. It has continuous operations and the work load depends on the customers that arrive and in the case of software maintenance, the work load depends on the Maintenance Work Requests (MWR) received. The activities carried out as part of a software maintenance are –

1. Receive Maintenance Work Request – this involves receiving the MWR, acknowledge it, record it in the MWR register which could be an excel sheet.
2. Analyze the MWR to determine what needs to be done, how to do it and the time needed to do it – The analysis includes determining the programs to be modified and estimation of the amount of time needed to modify those programs.
3. Plan and schedule the MWR – Planning would include deciding the priority for the MWR among the MWRs already on hand and then determining the dates for various activities like fixing the MWR, testing it, user acceptance testing and submitting it to the configuration manager who would update the software in the production server.
4. Allocate the MWR for implementation to a resource – This will involve identifying the person and handing the MWR over to that person for fixing the defects.
5. The assigned person completes the implementation of the MWR – The assigned individual implements the needed modifications in the identified programs.
6. The MWR is assigned to a tester to test and uncover defects, if any – the modified programs are assigned for testing to a tester for testing and ensuring that there are no defects in the modified programs.
7. Assign the uncovered defects for rectification – If the testing uncovers any defects, the modified programs are once again assigned to the person who originally implemented the MWR for rectifying the uncovered defects who would fix the defects in the programs.
8. Confirm the defect removal – Once the defects are fixed, the tester would be once again requested to test and ensure that the defects were indeed

removed and that the programs are without defects and can be subjected to user acceptance testing.

9. Conduct the user acceptance testing – The modified programs are tested by the originator of the MWR to ensure that the modifications asked were indeed implemented and that the programs are doing exactly what was intended.

10. Submit it to the configuration manager to update the product – Once the user accepts the modified programs, the programs would be submitted to the configuration manager. Configuration manager is in charge of the software on the production server and would ensure that all approvals are obtained and then update the production software on a schedule and update the software. Now on, the software on the production server would provide the changed functionality. This will complete the work of MWR and it will be closed.

Now, let us see what is involved in managing people in projects. Here are the activities that are performed from the standpoint of people management –

1. Allocate a project manager to take charge of project execution – The first activity in executing a project is the allocation of a software project manager (SPM) and handing over the purchase order, the project requirements and schedules asked for by the customer or the end-users. SPM would take all the responsibility to execute the project successfully.

2. Plan the project execution – SPM would carry out detailed planning and prepare the required plans which include the Project Management Plan, Configuration Management Plan, Quality Assurance (QA) Plan and detailed Project Schedule.

3. Prepare the work breakdown structure – Work Breakdown Structure (WBS) is a list of all the components to be developed to execute the project. It would usually be in an excel sheet or a project management software tool. It would have columns to record the name of the resource allocated, the dates of starting, ending, beginning of testing, completion of testing and so on. Work allocation to the resources would be carried out using this WBS. Once all the components are completed in all respects, the project would be concluded. The activities of integration at various levels also would be included as task in this WBS. An example of the WBS is given in Exhibit 6.2.

4. Raise resource requests as necessary – The SPM would estimate the resource requirements including the requirements of the human resources needed for the project. A sample human resource requirements

Exhibit 6.1   Human resource requirement form

Item Number	Type of Resource	Number Required	Required by Date	Probable Date of Release
1	Java Programmer	10	3-Mar-2019	3-Oct-2019
2	SQL Server Specialist	1	1-Feb-2019	3-Mar-2019
3	Dreamweaver programmer	2	1-Feb-2019	5-Apr-2019
4	Java testers	4	3-Apr-2019	3-Oct-2019
3	Java Team leads	2	1-Feb-2019	3-Nov-2019

Exhibit 6.2   Work breakdown example

Component Name	Module Name	Attribute	Estimated Size	Actual Size	Allocated for Construction to	Scheduled Start	Actual Start	Scheduled Completion	Actual Completion	Allocated for Review/Walkthrough to	Scheduled Start	Actual Start	Scheduled Completion	Actual Completion	Defects Uncovered	Allocated for Unit Testing to	Scheduled Start	Actual Start	Scheduled Completion	Actual Completion	Defects Uncovered

form is depicted in Exhibit 6.1. The resource allocation cell would use this resource request form to acquire the needed resources and allocate them to the project.

5. Allocate persons to the project – The resource allocation cell would allocate the resources as requested by the SPM making necessary adjustments in the allocated numbers, as necessary, for the skill levels of the allocated resources to give the SPM necessary number of resources such that the project can be executed on time as scheduled. Now the work begins for the resources. What were the allocated resources doing before allocation to a project? Obviously, they ought to be free doing no productive work and awaiting allocation. This goes to show that some amount of idleness is unavoidable in project-based organizations such as software development. The Resource allocation cell ensures that this idle time is minimized to the extent feasible, meticulously planning the release and allocation activities such that the gap between the dates of release and allocation is minimized.

6. Allocate the work to be performed periodically – Once the resources are allocated to the project, the SPM would conduct the project induction training covering the topics of the engineering tools and methods to be used in the project, modes of project communications, escalation mechanisms when necessary, configuration management and QA activities planned for the project and so on so that the project execution moves smoothly to completion. Then using the WBS mentioned earlier communicate it to the resource. Then the resource begins work and completes it and informs the SPM or the Project Leader (PL) as the case may be about the completion of the allocated work. We will deal this topic in greater detail in the next section as this is an important topic.

7. Take possession of the artifacts of the completed work – An artifact may be a code artifact (a program) or an information artifact (a document). Once the resource completes the assigned work, the SPM or the PL will take over the completed artifacts and check them in a cursory manner for completeness. Once they are satisfied with the artifact, the resource would be de-allocated and released to take up the next allocation. The artifact would be passed on to QA activities.

8. Subject the artifacts to quality assurance – The SPM or the PL would allocate the completed artifact to a QA person to carry out the planned QA activities on the completed artifact. The QA person would subject the artifact to a thorough review and testing as necessary to uncover defects lurking in the artifact. Once the QA activities are completed, the QA person would report the defects using a format or using an organizational software tool to the SPM.

9. Assign the artifacts in which defects were uncovered for fixing the defects – Once the QA activities are completed, and the defects report is available, SPM or the PL would assign the artifact for fixing the defects uncovered in the QA activities. Usually, this allocation would be to the same person who prepared the artifact and occasionally, it could be to another person if the original author is not in a position to take up this work. The person, to whom this activity is allocated, would fix the defects and report the completion to the SPM.

10. Assign the fixed defects to QA for defect closure – Upon fixing the defects, SPM would allocate the verification of the efficacy of defect fixing to the person who uncovered the defects or someone else, if the original person is unavailable. That individual would verify the defect fixing. If the defects were not comprehensively fixed or some new defects are uncovered, then the QA person would report the defects back

to the SPM. Until all defects are satisfactorily fixed, the artifact would go back and forth between the defect fixing and verification activities through the SPM. Once all defects are fixed, the QA person would give clearance for the artifact to be promoted to the next stage, which is usually the integration into a subassembly or the product itself, as the case may be.

11. Promote the artifacts that cleared QA for integration into subassemblies – The purpose of configuration management is to ensure that no artifact is allowcd to be delivered unless it passes through all the planned QA activities. For this purpose, the SPM would maintain separate folders for construction, that is initial development of the artifact, for QA, for integration, final build and then the delivery folder from where the delivery is effected. Once the initial artifact is completed, it would be moved from its initial folder into the folder for QA activities. Once it passes through its planned QA activities, the artifact would be moved into the integration folder. Once integration is completed, the subassembly would be moved into the QA folder for performing the QA activities' Once its planned QA activities are completed, it would be moved into the final build folder. Once final build is completed, the build would be moved into the QA folder for performing its QA activities. Once the build passes through its planned QA activities, it would be moved into the delivery folder. In this manner, the artifact is promoted after each QA activity is completed. This movement of the artifact from one folder to the other folder in order to prevent delivering an artifact that did not pass its planned QA activities is referred to as configuration management.

12. Go back to step 1 until all the components are built and passed the planned QA activities – The above steps are iterated until all the artifacts are built and put through their planned QA and one by one, all artifacts are promoted to the integration stage.

13. Assign the integration of components into subassemblies – Integration would not wait until all components are completed. Rather, when the components comprising one subassembly are completed that subassembly would be integrated. In fact, the SPM or the PL plans it so that the components are built in such a way that a subassembly can be integrated and tested. They achieve this using work allocation and the WBS. In this manner, all subassemblies would be built one by one.

14. Once integration is completed, subject the subassemblies to QA activities – When a subassembly is integrated, it would be allocated for carrying out its planned QA activities. The QA person would prepare

a defect report or report the defect using the organizational quality management tool.

15. Fix the defects uncovered in the QA of the subassemblies – The defects uncovered in the subassemblies could be in the integration part or the components. It is possible that some of the defects in the components cannot be detected until they are integrated. Once the defect report for the integrated subassembly is received the SPM or the PL would allocate the defects to one of the developers for fixing them. Once all the defects are fixed, they will once again be subjected to QA activities until all the defects are satisfactorily resolved. Once all the defects are closed, the subassembly would be promoted to the next stage in the planned configuration management for building the product.

16. Arrange for integrating the subassemblies into the product – Once all the subassemblies and other components needed for the product are available, the SPM or the PL would allocate the work of building the product. It would involve building any new components that are needed for the product; building interfaces between the subassemblies in the product; and then integrating all these into a product. It is likely to need more than one person. Once all this work is integrated, the product would be moved into the QA folder to subject it to the planned QA activities.

17. Once the product is built, arrange for QA activities – Once the product is built, it will be subjected to the planned QA activities. While the QA activities needed for components and subassemblies are a few, the product needs to be tested vigorously using a greater number of tests to ensure that the product will work without any hitch in the target environments. We will see more detail about product testing in the subsequent sections of this chapter.

18. Arrange for fixing the defects uncovered in the QA activities – As in the case of earlier components and subassemblies, the SPM or the PL would allocate the defects uncovered in the QA activities for fixing defects. The allocated persons would fix the defects.

19. Once all defects are fixed and assured, then arrange for acceptance testing by the customer – Once all the uncovered defects are fixed satisfactorily, they will be once again subjected to QA activities to ensure that the defects were indeed fixed satisfactorily. If any defects were not satisfactorily fixed or new defects are uncovered, they would go through another round of defect fixing and QA activities. Once the QA persons certify the product to be defect-free, it would be promoted to the next stage in the configuration management.

20. Once customer, who could be the end-users within our organization, accepts the product, deliver it and proceed with installation and commissioning – Once the product is ready and certified as defect free, there are two possible scenarios. One scenario is that the product is built for a specific single customer, which is referred to as the project scenario. The second scenario is to use the product as a COTS (Commercial Off The Shelf) product and sell to multiple customers. In the first scenario, the product would be offered to the customer for acceptance. The customer as part of acceptance, may test the product using their own testing methodology. If the customer uncovers any defects, they would be fixed by the project team. Once the customer is satisfied that the product is defect-free, they will sign-off their acceptance of the product. Then the SPM and the PL would move the product artifacts into the delivery folder in the configuration management. Then the delivery would be effected from that folder. Then, if it is included in the contract, the project team would go to the customer location and install the product on the target computer and perform all the activities necessary to roll out the product for production use by the end-users at the customer location. This completes the software development project for the specified software product.

6.4 Work Management

In managing people at work, managing work is the most important aspect especially from the standpoint of the individuals. As noted, persons come to our organization to perform the work assigned to them as a means to earning a livelihood. When work management is carried out effectively, the organization would have happy, productive, and stable workforce with high morale. The activities of work management are –

1. Allocation to the project
2. Induction into the project
3. Work allocation
4. Work de-allocation
5. Project-end performance appraisals
6. De-allocation from the project

Now let us discuss these activities in detail so we can manage work effectively.

6.5 Allocation to the Project

Since all software development work is carried out as projects, every time a new project is taken up by the organization, people need to be allocated to the project. Normally an organizational resource allocation cell (ORAC) or similar other department takes ownership of all the organizational human resources and fulfils the resource requests raised by the SPMs. One issue faced by the ORAC often is the matching of the skill set and the experience asked by the SPM with the available resources. More often than not, there is always a mismatch between what is asked for and what is available. While the resources with the requested skill and experience are available, they would be already allocated and working with other projects or on other similar assignment from which they cannot be immediately released.

So, the ORAC negotiates with the SPM to compromise on the skill set or the experience or both too in a few occasions! When the skill set is compromised, then project specific training to bridge the skill gap needs to be organized. If the experience is compromised, the productivity goals need to be scaled downwards. In such cases, ORAC allocates a few extra resources to compensate for the skill or experience.

It is pertinent here to note the two attributes of productivity. One is the level of skill and the other is the level of the effort put in by the resources on work. The Handbook of Industrial Engineering, edited by Maynard, defines five skill levels:

1. **Super skill** – a person who has extensive experience and who knows all there is to know about the work, including all the best practices and all the shortcuts. This person is an authority and guides others on the subject.
2. **Very good skill** – a person who has much experience and who has learned all there is to learn about the work, but who perhaps has not personally applied all of this knowledge. This person knows all the best practices and most of the shortcuts, and can guide others on the subject.
3. **Good skill** – a person who has adequate experience and is capable of delivering results, but who perhaps needs reference to documents or guidance from senior colleagues in advanced aspects of the work. This person does the work, but may not be able to guide others.
4. **Fair skill** – a person who is not a trainee and who has some experience, but who needs reference documents, or assistance from senior colleagues to complete the work. This person needs occasional guidance.

5. **Poor skill** – a person who is a new entrant and who has not completed the specified formal training period in the organization. This person needs continuous guidance from seniors.

The Handbook of Industrial Engineering defines five effort levels:

1. **Super effort** – not a second is lost, all available time is used; does not take any allowed breaks; no trial and error; all effort is focused on the task.
2. **Very good effort** – very high use of time; does not take all allowed breaks; no trial and error; all effort is focused on the task.
3. **Good effort** – uses time fully; takes only the allowed breaks; very little trial and error; all effort is focused on the task.
4. **Fair effort** – uses most of the available time; takes all allowed breaks; some trial and error; most of the effort is focused on the task.
5. **Poor effort** – uses as much of the available time as possible; takes all the allowed breaks and more; keen to experiment; all effort is not focused on the task.

When the SPM raises the resource request, the default skill level is the good skill. Of course, the SPM may request one or two super skilled resources if needed. But if the ORAC allocates all poor skilled resources even more numbers than needed, the project cannot be executed as there is no good skilled person in the team. So, in such cases, ORAC allocates a couple of people with more than good skill so that they can train and guide the poor skilled resources and get the project executed.

These attributes are usually considered by the ORAC while allocating people to the project. While the skill level can be known to the ORAC at the time of allocation, the effort that will be actually achievable is not known beforehand. It is something the SPM has to achieve. So, ORAC normalizes the allocated number based on the skill level and leaves it to the SPM to achieve the needed effort level. While planning the project beforehand is the key to success, the execution has a significant impact on the successful execution of the project.

While proper allocation of adequate numbers to the project is the responsibility of the ORAC, there is but one more responsibility which is equally, if not more important, of the ORAC. This is selecting the right resources to the project. When we say, right, we not only mean it from the standpoint of the project success but also from the standpoint of the allocated resource. A resource may not wish to work on a specific project for a number of reasons, which include:

1. The individual may have issues with the SPM or the PL. Conflicts are common to all organizations and software development organizations are no exception to this fact. If the resource and the SPM or the PL worked in a past project and had issues, the resource may resist allocation to the project. It is ideal that the resources do not have any feelings and will adjust with any other. But the fact is that software development is still a heavily human-dependent work.

2. The individual may not like to work on the development platform selected for the project. When the development platform is obsolete or is on the decline, software developers may not be willing to work on the project. Software maintenance on mainframe computers using COBOL language has a lot of people working. Still, software developers who are recent entrants would like to work on the latest crop of programming language which adds to their market value in the job market.

3. Some resources may not be willing to learn a new programming language for some reason like it is an obsolete one or is simply not interested in changing the existing expertise. This happens when a resource specializes in open source technologies and that individual is asked to change to Microsoft technologies, it would be a very big paradigm shift and the person may not be willing to work on that project.

4. The individual may not like to work on the domain in which the product would be developed. The domain, such as financial accounting or HR management, production management, may not be interesting to the individual because his/her specialty is in some other domain. For example, an engineer may not like to work of financial accounting and a person with accounting education may not like to work on production management.

5. The individual may not like work at the location from which the project would be executed. Most often, the project is executed at the location where the individual usually works but sometimes, it may be necessary to execute the project at another place such as the client location or at a geographically distant location which may upset the family life of the individual. Some people may be attending an evening school and shifting to another place may upset the education. In that manner, some resources can have problems in shifting to another location.

6. The resource might have worked with the client on an earlier project and had a forgettable experience and may not like to repeat the experience.

These are some of the important reasons why a specific resource does not want to work on a specific project. Allocating a resource to a project against his/her will would be detrimental to the project success besides adversely affecting the morale of the individual which in turn can adversely affect the team morale.

What should we do when a resource refuses/resists allocation to a project citing a good reason?

1. Find a win-win situation by finding a resolution that gives relief to the resource yet fulfils the project need. In this case, the resource sacrifices some of his/her comforts and the project takes a little discomfort. Both gain and both lose. The project gains the resources and the individual gains some of the need. The individual accommodates the project request and adjusts him/her self and the project gives the individual some concession.

2. We may try and locate a resource that is near enough fit for the resource in question but may need some training or guidance on the project. If such a resource is located, then we can allocate him and relieve the unwilling resource.

3. We may persuade the resource to accept the allocation for a short period until another equally competent resource can be located from the organizational resource pool. Most resources would accept if assured of guaranteed release from the project within a short, specified period.

4. If all the other resources are not fully utilized and some slack time is available in their allocation, then we may utilize such slack time and release the unwilling resource.

5. If all the other resources are fully utilized, we may have a temporary transient resource from a consulting organization and relieve the unwilling resource.

6. If there is absolutely no alternative to the unwilling resource, then we may have to persuade that person, by offering some incentive like allocating that person to a project of his/her choice or granting some vacation or some additional support and so on. Of course, we can use organizational authority, if required. But we do not advocate forcing any resource because it will have serious ramifications. It would send a message that the organization would use authority to force the resources to work on projects against their will. This is a serious side effect and we need to avoid this.

The above is a discussion on allocating resources to the project. But more frequent allocation is the work allocation. Once allocated, resources perform the work allocated to them by the SPM or the PL. This is an everyday activity. Resources are dissatisfied with the organization, more often than not, with the work allocated to them rather than organizational policies, or work environment or work methods. Most first line supervisors are ill-trained on the aspects of management especially on the matter of work allocation and de-allocation. They equate their role to policing, ensuring that the resources work without let up continuously for the 8 hours they spend in the organization. Policing is out of place in the organization today. Today's resources take responsibility, especially so, in software development as they all hold a minimum of a college degree. We must ensure that our first line managers are well trained in work management but alas, we find more of an exception to this than conformance.

The activities of work management at project level are–

1. Work allocation
2. Perform the work
3. Work de-allocation
4. Quality assurance
5. Project end appraisals
6. De-allocation from the project

Let us discuss them in detail here.

Work allocation – Work allocation to the resource is the most important of all other activities in work management. The work allocation involves, handing over the artifacts necessary for performing the work, the communication methods, means of obtaining clarifications when necessary, how to resolve roadblocks that may arise during the performance of work, escalation mechanisms when necessary and so on. Most of these aspects ought to have been covered in the project induction training but we may need to repeat them initially in the project. Once the going becomes smooth, we just have to hand over the artifacts to perform the work. The work needs to be allocated in such a manner that the resource should have the ability to perform the work successfully. In case where the resource is less than perfect for the task, then we need to support that individual by making available the guidance from an expert resource on the project on a need-basis. Then for each allocation we set some goals to the resources. These are:

1. Schedule goal – more specifically, the date by which the work should be completed. This needs to take into consideration, the date by which the artifact needs to be completed in all respects. We need to consider the amount of time that may be needed for the planned QA activities. We also need to consider that the possibility of the QA activities uncovering defects that may need rectification and one more iteration of QA activities. Therefore, the finish time set for the resource needs to be shorter than the final date set for the completion of the artifact in all respects. Then we have a responsibility to improve the capability of the resource by challenging and stretching his/her capabilities. This we can do by setting a slightly shorter target which will not unduly stress the resource but challenges that person. The pertinent question here is how much shorter should be the target that achieves these twin objectives? In our view, a 5% shorter target would not raise a red flag for the resource. We advocate a target that is 5% to a maximum of 10% shortening of the target date. This will raise the productivity of the resource.

2. Quality goal – It is never enough to somehow complete the work on time. It needs to be completed with the desired quality. Quality is measured in the number of defects per unit size of the component or the artifact developed. Program size can be measured in a number of ways but we suggest using the number of lines of code produced for measuring the size of the component. In documentation, we measure the number of defects per each page developed. We need to utilize the organizational baselines for this purpose.

3. Balancing work – We have a duty to ensure that all the resources in the project to be loaded uniformly with work. It does not augur well for the morale of the resources if some resources are heavily loaded and some are lightly loaded. We have to make a conscious effort to ensure that all the resources have similar amount of workload at any given point in time. Therefore, before allocating any piece of work, we need to check the work load on each of the resources and ensure that we load the resources with similar amount of workload and maintain parity in the amount of work performed by the resources. We can confidently assert that there is no other aspect of organizational working that contributes to both high and low morale besides the parity maintained in the workload distributed among the resources. We may naively expect that the resources do not discuss the respective workload among themselves but the fact remains that they do.

4. Frequency of allocating the work – This is also an important goal for allocating work. In project-based organizations, it is not possible to allocate work once without any need for reallocation for a long time, say, six months at a stretch or until a new batch of production is commenced. In project-based organizations, especially the software development organizations, we have the possibility to allocate work on a daily basis. The resources would despise it if they have to seek work every day. They would feel insecure if they do not know whether they have work allocated for the next day. We cannot allocate a very large chunk of work, a total subassembly, in one installment because we run the risk of too much pending work if the resource suffers a long absence. We need to divide the work into packages such that the inevitable absence of resources does not adversely impact the progress of work. We need to balance the frequency of work allocation in such a way, that the resource is secure about the continuation of work and at the same time, the project progress is also secure and immune from the unplanned absence of any resource. In our humble opinion, work allocation for a week at a time is well suited for software development environment. It is also better to allocate work on the last working day of the week for the coming week. This gives the much-needed security to the resources and they can enjoy the weekend without a care because the work allocation for the next week gives them the security and relieves them of any apprehension.

Perform the work – The resource needs to complete the work achieving the schedule and quality goals. While so doing, the resource needs to conform to the organizational decorum, improve upon the organizational baselines of productivity and quality. The resource also needs to assist other junior members of the project team. The resource also needs to foster team spirit and team play putting shoulder to shoulder and work in a close-knit manner for the overall success of the project and thereby add value to the organization.

Work de-allocation – What is there in de-allocating the work? Well, there is plenty. Both the SPM and the resource have to perform some activities. Here are the things to be accomplished. We need to note one thing though: it will not take days to do these things. It will all be done in an hour or two since some of it is already in place like the WBS, the allocation information and the targets. We need to do this every time a resource completes the work allocated to that individual. Here is what is involved in work de-Allocation:

1. The resource hands over the instruction artifacts, the developed artifact and any other information that is available with him/her. The developed artifact can be either a code artifact or an information artifact. Handing over would be in the folder defined in the configuration management plan.
2. The SPM or the PL takes charge of the artifact and moves it to the next folder defined in the configuration management plan.
3. The SPM updates the WBS with the completion information and computes the productivity and schedule metrics and compares them with the goals set at the time of allocating the work. If the variance, as there most certainly would be, is on the negative side, that is, the defects are more and the productivity is less, then SPM would analyze the reasons for the variance. If the reasons have known and unpreventable causes, then, no action would be taken. If there are no other reasons than the poor performance of the resource, then the SPM would counsel the resource on how to improve performance in the next allocation. Many times, this aspect is postponed till the end of the project just to avoid the unpleasantness. It is always better to counsel the resource as close to the time of occurrence as possible. The Hot-Stove theory of Douglas McGregor states that the counseling needs to happen as it happens in the case of touching a hot stove. When someone touches the hot stove, the burn is immediate and is commensurate with the hotness of the stove and the force of the touch. So, should our counseling be! On the other hand, if the performance variance is positive, that is, the productivity and quality achieved are better than the goals, then, we need to applaud the performance. We can do this in a team meeting or through an email or at a minimum, appreciate the resource in person. While human beings, and SPMs, are quick to criticize, we are shy to show appreciation. We need to do this to improve the morale of the resource. That will propel the resources to even greater performances.
4. The resource is allocated the next task to be performed, if there are pending activities to be performed. If there are no more pending activities appropriate for allocation to the resource, that resource needs to be de-allocated from the project and handed over to the ORAC for allocation to the next project. We will discuss the project de-allocation in the coming sections.

Quality assurance – Quality assurance of work is a big topic. Since this is not a book on software engineering, it is adequate to note a couple of points here.

Quality assurance comprises of two functions. First is the set of activities focused on preventing defects in the deliverables and the second set of activities is focused on uncovering defects from the artifacts being developed. The first set is referred to as defect prevention activities and the second set is referred to as quality control (QC) activities. Review of the artifacts by peers and testing the artifact are the two vital quality control activities. These two are also referred to as verification and validation activities. Whenever a resource declares that an artifact is completed and is fit to be subjected to QC activities, the SPM would allocate the artifact to a person expert in the QC activities. Information artifacts cannot be subjected to testing and are subjected only to peer review. Code artifacts, on the other hand are amenable to both review and testing. So, they would be subjected to both the review and testing activities. So, when the artifact is information artifact, the SPM would allocate it for peer review and arranges rectification of the defects uncovered. There ends, the QC activities for the information artifacts. For code artifacts, the SPM would first allocate the artifact to the peer review and arranges for fixing the defects. When all the defects uncovered in the peer review are closed, only then will the artifact be allocated for testing. Testing is likely to uncover defects even when all the defects uncovered in the peer review are fixed. So, the SPM would subject a code artifact to both the QC activities and ensures that all the defects uncovered in both the QC activities are fixed. Until both the QC activities including fixing and closing the defects are completed, the artifact would not be promoted to the next stage in the configuration management system.

De-allocation from the project – Deallocation from the project is a much bigger activity than de-allocation from a task. It is likely to take a day or two. Here are the activities that are carried out as part of de-allocating a person from a project:

1. Carry out the performance appraisal for the employee and after agreement with the resource, handover the performance appraisal to the HR department.
2. Update the skill database of the organization with the new skills acquired by the resource during the course of project execution.
3. Takeover all project related artifacts including source code, code libraries and development kit.
4. Archive the project related communications that may need preservation as part of project records for future reference and possible use (note: this is a legal requirement in some countries).

5. Prepare the project release communication (a formal letter or an email) and handover the person to ORAC.
6. Document best practices, worst practices or any events deserving special mention about the resource and forward the same to the concerned management personnel.
7. Takeover the software artifacts, communications and hardware resources from the employee to concerned agencies in the organization for further maintenance and re-allocation.
8. Update the project information to reflect the release of the resource.

Once all these activities are completed, the resource is released from the project and is deputed back to the ORAC.

Project end performance appraisals

Performance appraisal is big enough topic to deserve a separate chapter, we deal with it in detail in a subsequent Chapter 10. Here, we just emphasize the fact that at the end of every project, we need to carry out a performance appraisal and include it in the personal records of the resources. We limit the performance to the current project only. The project end appraisal would need to be a subset of the year-end performance appraisals. The project-end appraisal would have lesser aspects than the year-end appraisal and the detail that needs to be included ought to be much less than that of the year-end appraisals. Why do we need to have a project-end appraisal when we would have a year-end appraisal? Here are the reasons –

1. If the resource works with only one manager, the manager would have the complete record of the employee performance spread over the entire year. Therefore, that manager can competently and with full information carry out the performance appraisal once a year. In software development organization, the resource would be working on multiple projects in a year with multiple managers.
2. The resource may in all probability, work on multiple projects, managed by multiple SPMs in the year and no single SPM would have the complete record of the employee performance spread over the year to competently carry out the performance appraisal.
3. When the year-end performance appraisal becomes due, all the SPMs under whom the resource worked may not be available to meet and carry out the performance appraisal of the employee. Second, it is impractical to arrange the meeting of all the concerned SPMs for each resource to sit together and carry out the year-end performance appraisal. If there are

1000 employees, we may need to have 1000 separate meetings to complete the performance appraisals for the organization! It is impractical and naïve to attempt such a process.

4. With project-end appraisals, the HR department itself can compile all the project-end appraisals and finalize the year-end performance appraisals for all the employees quickly and effectively.

That is why, we need to carry out the project-end appraisals whenever a resource is being released from the project. It ought to be the last task of the resource to participate in his/her performance appraisal to take stock of his/her project performance. Once the project-end performance appraisal is signed off, the resource can leave the project with information about the areas that need improvement in the next project and the strength areas that the individual can continue to bank upon.

6.6 Goals of Work Management

Work management activity has four goals to achieve. These are –

1. Work completion – The main objective of work management is to produce a deliverable product adhering to its design and organizational quality standards. The product needs to be completed. At the end of the project, an integrated, functioning and deliverable product should be ready for delivery. A project should not end up with disparate components that are either incomplete or unintegrated into a product. Having a fully functioning product is the primary objective of managing work.

2. Quality achievement – We have to complete the product but the product should have built-in quality! The quality of the completed product needs to conform to its specifications fully. The specifications might be national, industry standard or organizational standard. Whatever standard was selected for our product, the completed product ought to adhere to that standard in all respects. Quality gets the organization repeat orders and a reputation in the market. The second most important objective of the work management is to achieve the specified level of quality for the deliverable.

3. Productivity management – Productivity is the rate of achievement. IFPUG (International Function Point User Group) defines productivity as the Project Delivery Rate (PDR). Both mean more or less the same. It is specified as the amount of work performed per a unit amount of

time. In programming, it is usually specified as LOC/Day. LOC is the number of lines of code. Day is one working day. Some people use other measures like the Function Point, Object Point or Software Size Unit. Each type of work would have a productivity set for the product. We need to ensure that each resource on the project achieve the productivity set for the project or improve upon the norm. If we do not achieve the productivity norm set for the product, the profit derived from the project would become less. Higher productivity means higher profit and lower productivity means less profit. As organizations exist primarily to make money for the investors, we need to manage productivity so that we earn adequate profit for the organization and the investors.

4. Morale management – We need not overemphasize the importance of morale in organizations. Teams with higher morale can beat the world. Teams with lower morale cannot cross the road! We have a duty to maintain, if not improve the morale of the team as a whole and the team members individually. The SPM has a duty to ensure that the team morale is maintained and improved. The topic of morale and motivation would be dealt in the subsequent Chapter 9 in greater detail.

6.7 Conclusion

Work management is of paramount importance in the organizations in general and software development organizations in particular. Alas, we do not accord adequate importance to this aspect in the organization. We leave the aspect of work management to the first line supervisors. It needs to be overseen by senior management. Whether the organization would be a winner of loser is decided by work management alone. We need to accord more importance to this aspect and ensure that it is managed well.

7

Employee Relations

7.1 Introduction

Keeping the employees happy and satisfied, as a prerequisite to getting the desired productivity and quality work from them, requires a set of policies and practices that ensure uniform and fair treatment of the individuals employed in the organization. The onus is on the organizational management to set those policies. Irrespective of the supervisor with whom the resource works, the treatment meted out to the individual needs to be the same. When unions stepped into the organizations, a department was needed to represent the organization with those unions. That department needed to study the labor law and ensure implementation of the same in the organization. It interfaced with the both organizational management and the unions, working as the shock absorber, to maintain a harmonious environment in the organization amenable to ensure continuous productive operations. Since unions were in the industries, these departments came to be called as industrial relations (IR) departments. In industries such as software development where there are no blue-collar workers or workers unions, the IR department is conspicuously absent. Even though the ILO (International Labor Organization) brought out the compendium on professional workers in 1983, which allowed the right to form unions for professional workers, they did not take to unionism in a big way. While the need for interfacing between the management and the union diminished, there still remained a need to implement labor laws related to professional workers, to ensure absence of any sort of discrimination and harassment in the organization; a department however small it might be, was needed in the organization. While there is no uniformity in naming this department, we prefer to use the phrase "employee relations" for want of a better title. In some small to medium organizations, there would be just one executive looking after this function and in small organizations, the head of HR would assume this responsibility. In this chapter, let us discuss the roles and responsibilities of this function.

Noted psychologist Frederick Hertzberg developed two-factor theory for motivation. While he developed this theory in 1959, we feel that it is relevant even today. Managements of organizations did take a serious note of this theory and implemented it in the organizations. Today, all organizations either explicitly or implicitly implemented this theory. He named the two factors as Hygiene Factors and Motivators, as critical to higher levels of motivation. Hygiene factors are the salary, fringe benefits, working conditions and other non-financial benefits fall under this category. Basically, hygiene factors are those that define the organization. Then motivators are work itself, recognition, opportunities to advance knowledge, good supervision, involvement in matters that concern the employee and so on. Motivators are implemented by the concerned supervisors. While hygiene factors are part of organizational framework, motivators are part of all levels of supervisors. We discussed supervisors' role in ensuring presence of motivators in Chapter 6 on work management. Now let us discuss the organizational framework that ensures the hygiene factors in this chapter.

One thing of importance to note in this chapter is the age-old 90%–10% law! That is 90% of the cases, we follow a single rule but in the remaining 10% cases, we need many more methods and processes!

7.2 Functions of Employee Relations

The employee relations department under whatever name it operates performs some functions vital for the organization. These are:

1. Organizational framework
2. Salary administration
3. Pay hikes
4. Career management
5. Ensure compliance to statutory framework
6. Grievance handling

Let us now discuss all these functions in detail.

7.3 Organizational Framework

The main objective of the employee relations function is to keep the employees satisfied and fulfill the organizational responsibility in that direction. For the employees to be satisfied and maintain a high morale, two aspects are vital. One is the workplace where the employees spend all their work life.

Fair and equitable treatment by the concerned supervisors is needed to ensure employee satisfaction. The second aspect is the one that is the organizational framework within which the workplace supervisors operate. This needs to be defined and maintained by the organizational management and the function of employee relations assists the management to that extent. What does the organizational framework for keeping the employees satisfied consist of? The organizational framework for maintaining harmonious employee relations consist of:

1. A set of policies for keeping the working environment pleasant. A few years earlier, "work environment" meant an office where all the employees came to spend the next 8 hours doing their work together. Now, quite a few offices closed shop and asked the employees to work from home. The spread of internet made it possible for people to connect with the organizational computing resources and carry out their work. The software development work is especially amenable for this sort of arrangement and they adopted it in a big way. This arrangement is convenient to the organizations as they can cut costs of maintaining an office including rent, energy charges, communication costs, house keeping costs and so on. For the employees, they saved traveling time, work from the comfort of their homes, sans the official/formal dress, have flexibility in the working hours and so on. Now, in this environment, what does the organization do to maintain a pleasant work environment conducive to higher productivity? A good question, which we need to answer here. We can have most of our resources working from home but not all. We need a smaller office to have some staff to coordinate the activities of those working from home, arrange meetings, both face-to-face and online meetings as and when necessary, maintain infrastructure that is essential to facilitate from-home-working and such other activities. Then again, we need to have mechanisms to ensure that the people working from home are really working and turning in acceptable productivity and quality. All these come under the work environment.

2. Whatever work environment we may choose, we need to have a set of defined policies, procedures, formats and templates, issue resolution/escalation mechanisms and so on in place so that work can be performed as smoothly as possible. This is also part of the organizational framework.

3. Continuous improvement of the organizational framework to cope with the changes that inevitably take place is essential for the work environment to be effective. This is also part of organizational framework.

4. Finally arranging for the facilities needed by the project teams for effectively executing the project including the facilities for internal testing as well as customer testing and so on.

All these form part of the organizational framework utilizing which the project teams can effectively execute the projects.

7.4 Salary Administration

Salary administration is a large enough subject and it deserves a separate chapter. Therefore, it will be described in detail in Chapter 8. Here we touched upon the subject because the onus of diligently administering the salaries of employees, rests on the shoulders of the employee relations department.

7.5 Administration of Leave of Absence

The second and most important aspect of salary administration is to pay it, regularly and without mistakes. After all, people come and work in organizations to earn a livelihood and salary is the crucial component of that livelihood. The second aspect of payment of salary is the administration of the leaves-of-absence (LOA) allowed by the organization. We find various practices in the LOA across different organizations and countries. In some countries it is a practice to allow LOA as a fixed number of days per year. An employee may avail a LOA for any purpose. An LOA is treated as a privilege of the employee which can be availed anytime by the employee. For short durations, usually for a day or two at the most, LOA is usually not denied but may expect the employee to inform in advance. But if the employee wishes to take LOA for longer durations, it needs to be planned such that the absence would not hamper the progress of the project adversely. LOA for longer durations, more than three days at a stretch need to be availed after agreement between the resource and the supervising executive. Sometimes, longer LOA becomes inevitable for some important occasions and emergencies. Suppose, the parent of the employee falls sick; or the marriage of a sibling is fixed at short notice; or the employee may fall sick or meets with an accident. In such cases, the LOA should not be denied.

Then, in some organizations, LOA are classified into three categories. The popular classification is to categorize LOA into casual leave, sick leave and paid vacation. Casual leave is that LOA taken for short duration of one or two days at the most. Sick leave is permitted only when the employee needs medical treatment for sickness or injuries require the employee to rest at home or hospital. In some cases, the sick leave would not receive full salary and receive a certain percentage of the regular salary. Paid vacation is such LOA that can be availed for longer durations. Different companies and countries would have different rules for this kind of leave. It is common to allow two weeks paid vacation per year in USA, but it is longer in countries like the France. Usually, companies that also pay leave travel concession club this vacation with that allowance also. Paid vacation is allowed once a year. Some organizations allow the employees to accumulate the LOA other than the sick leave by carrying forward the un-availed LOA to the next year. That way, the employees can enjoy longer vacations.

In many instances in most organizations, some employees need to work extra hours on a working day or work on a holiday due to exigencies of work. Some organizations pay extra pay usually at double the usual rate for this extra work. This is normally referred to as overtime pay. Some organizations follow a practice of giving compensatory off (referred to as "comp off" at some places) to compensate for the extra work. While the overtime pay is at double the usual rate, compensatory off is usually a day for a day. That is, if a resource accumulates 8 hours of overtime working, that individual would be allowed to take one day of comp off. Comp off would be restricted to a minimum of half a day at a time.

Most organizations allow another class of LOA generally referred to as special LOA. Sometimes, the employee needs to take a longer LOA than the one allowed to his/her credit including the carried forward LOA of previous years. The occasions that necessitate that kind of leave is a long bout of illness to self or a close family member. This special LOA is treated differently in different organizations. Even in the same organization, the treatment may differ from case to case. In most cases of special LOA, no salary is paid for the special LOA but in some cases, such as a critical employee or a loyal employee, salary may be paid as an advance to be adjusted in future salary payments in installments. In some cases, the special LOA may be adjusted against future accruals of LOA. In some rare instances, the management may even pay full salary without deducting it from future payments at its discretion.

All organizations that employs female employees do provide maternity LOA for their female employees. In earlier days, it was allowed only to married female employees but now it its allowed to all female employees. It is also normal to provide this LOA for abortions too. But there would be restrictions on the number of occasions this LOA can be availed. It is rare for organizations not to place any restriction on maternity leave as it would put a significant financial burden on the organization as the maternity leave is for much longer periods than even the yearly paid vacation. Generally, the LOA for giving birth to a child is twice as long as the LOA allowed for abortions. Now, many progressive organizations have recognized the need of male employees to be at the side of their spouse during time of delivery and allow a paternal LOA to males who are becoming fathers. Now, some organizations insist that the man availing paternal LOA needs to be the husband of the female giving birth but some do not place any such restriction except some sort of proof that he is the father of the child.

There are also cases when resources absent themselves from work without obtaining any LOA for long durations at a stretch. Obtaining authorization for LOA indicates that the employee would return to work after the LOA. But if the employee absents from the work without any intimation, what should the organization do? Generally speaking, the organization implements these steps when some employee is absent without authorization for long periods.

1. Each organization has specific duration to raise an alarm because, the employee may be in some sort of medical or other emergency and is unable to communicate with the organization in spite of his/her best efforts. Once this threshold is crossed, the supervising officer of the department/project in which the absent employee is positioned currently raises an alarm. This alarm is a communication to the employee relations department about the absent employee.
2. The employee relations department attempts to establish communication with the absenting employee by phone or email or contacting the employee's relatives and friends, and finally paying a personal visit to employee's address in the records by an authorized person. Unless all other attempts fail, personal visits would not be resorted to as it is perceived as breach of the privacy of the employee.
3. In extreme conditions, an advertisement may also be placed in an appropriate medium such as TV. This would be utilized in special cases like natural calamities.
4. When the attempts to contact the absenting employee fails, the company places such employee under suspension, that is the employee is still retained on the organizational rolls but would not be paid any salary or a small portion of salary becomes due to the employee.
5. Some organizations keep this suspension eternally. But most organizations keep the employee in suspension for a specific period like a month

or so and then terminate the services of the employee for unauthorized absence. There could be legal implications for keeping the employee in suspension for longer periods. Suppose, if the employee is discovered to be expired during the period of suspension, the organization may become liable for compensatory payments because the employee is technically still on the rolls. Every organizations needs to examine the applicable law and set these limits.

6. Once the threshold for keeping the employee on rolls under suspension is crossed, the organization terminates that employee. We at employee relations need to implement the policy defined for such cases. We need to intimate the finance to cease payment of salary; the concerned department that deals with separations to take necessary actions to terminate the services of the absenting employee; the recruitment department to recruit a replacement for the absenting employee.

Employee relations department needs to handle all these actions as part of administering the LOA of the employees.

7.6 Maintenance of Leave Records

LOA is treated as a right of the employee. LOA is a very important record from the standpoint of an employee as the employees plan their vacations well in advance. Any discrepancy in the balances would upset the employees severely. Leave records may not be able to increase the level of motivation of the employees but errors thereof could severely affect the employee morale. Therefore, we need to maintain the leave records of the employee meticulously in such a manner, that the employee can get the information about the balance available, the actual instances of taking LOA, as well as reconcile his/her record with that of the organization. Now, it has become normal to maintain all leave records on computer. Here are the actions performed by us in this aspect.

1. Individual leave record for each employee on the rolls needs to be maintained. It will consist of all the LOA availed by that individual. The tenure of this record is determined by the record-retention policy of the organization which usually would be for the last five years. This record will capture the details of the dates on which the LOA was availed, the type of LOA, who approved it, any extensions were availed and if yes, for how long and any other organization specific details would be captured.

2. At every instance of availing LOA, the following actions would be taken by the employee relations personnel:

   (a) Capture the details of LOA in the employee leave record.
   (b) Reduce duration of the LOA availed from the LOA balance standing to the credit of the employee and update the balance LOA available to the employee for future consumption.
   (c) At the beginning of the year, which is usually the financial year, credit the LOA account of each employee with the LOA that becomes due to the employee for the coming year and update the balance of LOA standing to the credit of the employee.
   (d) If the employee avails special LOA, either with or without pay, then record details of payments made to the employee or deducted from the employee salary for the special LOA period and record it in the leave record of the employee.
   (e) In case any employee becomes absent without proper authorization, keep track of such absence and initiate actions as and when necessary.
   (f) Some organizations provide cash reimbursement against LOA balance standing to the credit of the employees at the request of the employees. In such organizations, there would be a set of guidelines as to the percentage of LOA standing to the credit of the employee that can be surrendered for cash by the employee, the dates when the encashment can be availed, the formula for computing the money that the employee becomes eligible to receive on surrendering one day of LOA and so on. Whenever any employee opts for encashment, the employee relations department would scrutinize the application for completeness, eligibility and other details for accuracy and then passes it on to the finance department for releasing the payment. The leave records of the employee would be updated by reducing the balance LOA and the details of LOA encashment which will be checked in the next iteration of LOA encashment initiated by the employee.
   (g) Prepare and provide periodic reports, normally on a monthly basis, to the management about the total LOA in person days availed by the employees, the number of employees that availed LOA, total encashment of LOA availed by the employees, if any, and the money paid for such encashment. This report will give an idea about how many person days were lost to the organization or how

many person days were productively used by the organization in carrying out the work under progress. It will also provide reasons for any delayed deliveries or early deliveries and so on. If case excessive LOA was availed by the employees, it may also give a pointer to the management about any lingering dissatisfaction among employees that resulted in LOA en-masse. It allows them to investigate the issue and take corrective and preventive actions.

Now, it is the employee relations department that needs to administer the LOA by performing all the above-mentioned activities. We keep proper records of the LOA policies, various LOA taken by each employee, arrange payments allowed for the LOA taken by the employee and take actions necessary for special LOA as well as the actions needed for LOA taken without authorization.

7.7 Career Management

Career advancement is a component of motivating employees. Actually, there are four levels in the organization. The first level is the working level where the productive work is carried out. The employees in this level perform work that result in the deliverable. The second level is the first line supervisory level. The employees in this level closely supervise the work being carried out at the working level. These people assign the work to the working level employees, guide them in performing the work, provide expert assistance when needed and in general, facilitate the working level employees to perform their work without any distractions. The next level is the managerial level. The employees in this level plan, organize and staff the workplace to ensure that work is carried out smoothly without any hindrances. They ensure that the effort put in by the working and supervising levels is integrated into a working deliverable; the work is carried out at the desired level of productivity ensuring planned profit for the organization; and ensure that the product is built with the desired level of quality. They coordinate between various functional specialties to ensure that they work in a close-knit manner. The highest level is the senior management level which consist of profit center heads or departmental heads in large organizations. These people take ownership for the profitable functioning of one independent entity. The top management is the owners, investors, board of directors and so on. These are not just employees; they are more like owners. Large organizations may employ professionals in these positions, but their career need not be managed.

When we talk of career management, we speak mostly of the bottom three levels, namely, the working level, the supervising level and the managerial level. The senior managerial level people are at the top of the heap and further rise can happen only based on organizational need as determined by the top management. Recruitment to these levels would be based on the educational qualifications besides experience. The qualifications needed for the supervisory level would be superior than those required for the working level. Similarly, the qualifications needed for managerial level would be superior to those of the supervisory level. Senior management level would demand great and proven track record in managerial level.

While the levels are just three, there would be multiple layers in each level. To provide motivation to the employees in these levels, it is common to promote them once every two or three years. The shifting from one level to the next higher level is a usually a big event preceded by some sort of testing to ensure that the person is indeed fit for the next level. That can happen only when the person in the lower level reached the top layer in the level and put in some years before he/she can be considered for the next level. How many layers ought to be there in each level? In construction and manufacturing organizations, it is common to have about five layers in each level. The climbing the ladder from the first layer to the top layer can take from ten years to 15 years. That is, a person joining at the bottom most layer, diligently works and gets the promotion to the next layer once every two or three years, can reach the top layer in ten years at a minimum of 10 to 15 years at the most. In that time, the person becomes master of the trade (s)he practiced and would be ready for shouldering the responsibilities of the next level.

If we do not provide any promotion for a period of ten years even though we give great pay hikes every year, employees would not be motivated to put in better performance. That is why once every two or three years, a promotion to the next layer is given to the employees so that they do not feel stagnated. Manufacturing organizations are hierarchical in nature with pyramidal structures, and so they have about 5 per level. Service industries including software development organizations on the other hand are not that pyramidal in nature and have flat structures. Therefore, they normally have about 3 layers at each level. The elevation from one level takes place based on these criteria:

1. The person would be in the final layer after which there is no further layer and career advancement needs to be to the next level.

2. The individual possesses the qualifications specified for the next level.
3. There is a position available for the candidate to hold.
4. The elevation from one level to another is cleared usually after an interview and the candidate cleared such an interview.

The elevation from managerial level to senior management level is not automatic. The elevation to senior management level happens only when a senior management person leaves the organization or when a new department or profit center is created. Often, top management prefers to recruit outsiders to this kind position because, there may be resistance from within the organization for the elevation from inside resources. That is why, people stagnate at managerial level by staying in the same company for a long time. There is more attrition at the top layer in the managerial level, especially from the senior people in that layer.

The elevation from one layer to the other is normally automatic based on a time scale. That is, one a person puts in the required number of years in one level, that individual would be promoted to the next layer automatically. This kind of promotion orders would be issued along with pay hikes after the expiry of the organizational financial year. Even while these promotions are timescale-based, the performance appraisals would be taken into account, for giving the promotions. For those employees whose performance is above an acceptable performance rating, would receive the promotion.

What about those employees whose performance rating falls below the acceptable rating? In some cases, their promotions would be deferred by one year. If the performance rating bounces back, then the person would be promoted. Such employees lose just one year and fall behind their peers by that much time. In some cases, especially when the performance rating has been consistently rated below the acceptable level, at least for the previous two consecutive years, the individual may be placed in some sort of performance improvement plan. The decision of placing a person on performance improvement plan is taken by the employee relations department based on the performance appraisals of the previous years in consultation with the departmental head in which the identified employee happens to be working. We need to ascertain the counseling administered to the employee during the period in which the performance was below acceptable level. That is, we need to ascertain that the employee has been given ample opportunity to improve himself or herself. Only when we become certain that the employee is the reason for the unacceptable level of performance, we place an employee on performance improvement plan.

The performance improvement plan has the following components:

1. Recording the present level of performance that has been accepted by the employee and the executive supervising the employee.

2. Setting measurable goals of performance improvement – These goals need to be agreed to by the employee. While so, the set goals must be such that the value addition by the employee will be equal to the acceptable level of performance, at a minimum. We cannot set goals below the acceptable level of performance, even though they are better than the existing level of performance of the employee, just because, the employee accepts those goals. What we mean to say is the employee acceptance of the goals is not the sole criteria for setting the performance improvement goals.

3. Identifying the milestones in the performance improvement so that the progress can be objectively assessed when the employee arrives at the milestone. When the employee arrives at the final milestone, the performance improvement plan is completed successfully.

4. Measurement of performance level at preset regular intervals is important to assess the progress. While arriving at a milestone clearly establishes the progress, we should not wait for the employee to arrive at the milestone. We need to measure the progress at preset intervals to measure the progress and contrast it with the preset objectives of the progress. This helps us to effect any needed course correction, if necessary.

5. Counseling and coaching after every measurement cycle helps the employee to sustain the pace of improvement or improve the pace of improvement, should the achievement fall short of expectation. Counseling is to bring the progress to the notice off the employee with emphasis on the direction of the progress. Coaching is to assist the employee to improve the performance of the employee.

At the end of the performance improvement plan period, we need to take a decision about the future of the employee. Three outcomes are possible.

1. The employee improves the performance as expected so that the employee can be continued in employment carrying out the duties assigned to him/her in the normal manner. The employee relations department would not concern itself with the employee any further.

2. The employee does not improve as expected which can result in two possible courses of actions. One, the employee may be discharged from the employment and sent out of the organization. This happens when the employee performance is so dismal that further efforts on improving

the performance are viewed as futile and pointless. Two, the performance improvement plan may be extended for some more time. This happens when the employee improves but the improvement is perceived to be inadequate but the potential for improvement is perceived to be significant.

In this manner, we manage the career of the employees. When an employee joins our organization at the bottom layer of a level, it would take that person, 10–15 years to rise to the next level. If the employee joined our organization at the bottom level in the bottom layer, straight out of college, it would take that individual about 30 years to rise to the top layer of the managerial level. As we noted earlier in this section, rising to the senior management level depends on the need of the organization and the ability exhibited by the person.

Of course, the employee may face speed bumps in elevation from layer to layer and level to level. In that case, the individual takes more time to reach the top layer of the managerial level. It is also possible that the individual can be promoted ahead of his/her time if the results achieved are much better than expected and are clearly visible and notable. In such cases, the individual takes less time to reach the top layer of the managerial level.

We need to understand that there are super performers in our employees. It would be naïve to expect all the employees to be at the same level of performance. Some would perform superbly right from the beginning and consistently. We need to identify such super performers and encourage them. The performance appraisals form the basis for this identification. When the performance appraisal of a specific individual is rated very high, then, we at the employee relations department need to note this performance and keep a watch on that person. When two consecutive performance appraisals of a specific employee are rated very high, we need to bring this fact to the senior and top management to test that person out by giving tougher assignments to ensure that the ratings were not fluke but were real. When the senior management pushes the limits of that employee, such persons may be earmarked for special treatment and be put on fast track career movement.

Of course, it does not mean that once an employee is placed on the fast track, we forget about that person. We continue to monitor the performance. It is possible in some rare cases that an employee in the fast track to be placed on performance improvement plan. Once a person comes into performance improvement plan, he could return to the normal track once he/she is off the performance improvement plan. Would it be possible for the person placed on performance improvement plan to come on the fast track? Why not, but it would need some strenuous effort from the individual to demonstrate superb results and much more than expected performance. Once the employee returns to the top level of performance, he can be put back on fast track. Once an employee, we should not demonstrate any bias against any employee except for the performance of individual on the job.

7.8 Ensure Compliance to Statutory Framework

Every country had enacted laws in every aspect of life including how to run an organization with regards to safety of the people working in that organization, working hours in a day, public holidays and weekly off, wages and salaries, LOA, building code and other laws. There are stringent laws pertaining to running a public company and handling of finances. For these two aspects special officers are appointed and they look after those affairs. If the company raised money from the public, then other statutory agencies come into the picture asking questions and demanding compliance with the laws of shares and stocks, dividends, debentures and related aspects. The companies have specially designated officers to handle those compliance issues. We need to remember that a company is a legal entity which conducts a business, pays taxes, employs people and is subject to the laws of the land. From the standpoint of the employee relations department, we have the onus of implementing laws concerning the human resources and ensuring compliance to those laws in our organization.

The following aspects of compliance to applicable laws is the responsibility of the employee relations department.

1. Working conditions including provision of common facilities such as toilets, change rooms, canteen, creche etc. We need to maintain a clean and pleasant environment which keeps the employees safe and secure on the organizational premises.
2. Access to emergency medical care or at least a first aid facility. We need to keep a first aid kit at convenient positions so that it is available to employees when needed. We should also train all or some persons with the skill to administer first aid to employees when needed before they can be shifted to a medical facility.
3. If women work in our organization, there are separate statutes in most countries to ensure proper working conditions for females. We need to ensure these facilities for them:

(a) Creche for small children on our premises or close by in the vicinity, especially in factories or such other buildings.

(b) Special LOA for maternity purpose.

(c) Special facilities for toilets and changing rooms.

(d) Safeguards from sexual harassment and a redressal mechanism for actual occurrence of such events.

4. Working hours adhering to the statutory laws and overtime pay for extra hours of working or compensatory off thereof.

5. Enforcement of weekly off during weekend as applicable in the country.

6. Holidays when the entire organization is closed including special pay for the emergency staff that need to work on the holidays. Some of these holidays must coincide with the national holidays when the whole country would have a holiday, on such days as the Independence Day.

For all these aspects, we issue internal SOPs (standard operating procedures) and appoint executives to oversee these aspects and resolve any grievance. Then we need to ensure safety of employees working in our organization while they are on the premises. This includes accidents caused by any reason including accidents due to natural forces like hurricane winds or a truck ramming against the walls or our building or an aircraft landing on us! Of course, we need to take an insurance policy to protect our organization and the employees from such unusual occurrences. For other workplace safety, we need to provide for firefighting equipment for small accidental fires like the fire extinguishers, first aid kits, protective headgear and other safety equipment depending on the work carried out by the specific employee. For example, an electrician is given insulating gloves and tools to protect him/her from electric shocks, gas masks for those that need to work in polluted environments and so on. On the security front, we need to ensure the following:

1. Ensure a working environment that is free from any pollutants and odors. Where odors are a byproduct of the deliverable like in a kitchen, we need to provide exhaust facilities so that the odors produced are driven out. We need to ensure that the workplace is dusted regularly to ensure a dust free environment.

2. We need to ensure that slippery substances are not spread on floors to prevent injuries to our employees due to falls caused by slippery floors.

3. We need to regularly clean our floors to ensure clean floors and to remove any items that fell on the floor due to oversight of people in the vicinity.

4. We need to ensure that all fixtures of electrical and other equipment are securely fastened so that they do not cause injury to our employees.

5. We need to ensure that no lose parts or any other loose items are stored recklessly at a height that may fall on the employees.

6. Provide all safety equipment and gear to employees that are essential during the working hours. We should also keep safety equipment like the fire extinguishers close at hand for emergencies.

7. We need to put in place emergency procedures like evacuation in the event of a fire accident or an earthquake or some such other calamity.

8. We need to regularly clean our common facilities like the toilets, water fountains, coffee machines, and canteens if any so that good hygiene is maintained.

9. If there are any hazardous material on our premises, we need to ensure that they are kept away from naked flames to prevent fire accidents. Similarly, smoking needs to be restricted to designated areas to prevent fire accidents. In software development organizations, there is likely to be a lot of wasted paper which, if not handled properly, can cause a fire accident.

10. We need to train our employees for using all such safety and emergency equipment so that they are ready for any eventuality.

11. Any other relevant aspect of organizational working that can cause injury or damage to either premises or the employees need to be handled by us.

Besides, putting in the required facilities to ensure implementation of the applicable statutes, we need to put in place systems and procedures to record the implementation and submit reports to the management and the statutory authorities. We need to keep a record of all the above aspects including periodic inspections to ensure that the facilities are maintained as needed and that the safety equipment like the fire extinguishers are in working condition and prepare inspection reports which are approved and filed in records.

Audits are the main tool to ensure quality of these services. Audits are periodic document verification systems to ensure compliance of the practices to the precepts. They are carried out once in a quarter or so by peers. Each function is audited for about an hour or so verifying the reports and other documents to ensure that all procedures are implemented diligently. If audit finds an NC (non-compliance), that is the applicable procedure was either not implemented totally or not implemented as it should be, the auditor would raise an NCR (non-compliance report). Before the next cycle of audit, the auditee needs to correct it by taking an immediate corrective action and a preventive action to prevent the recurrence of the NC in future. In each audit cycle, all the NCRs are compiled and presented to the management to draw their attention to the issues so that they ensure providing resources and monitoring the area.

We are usually required to submit such reports to the local statutory authorities to assure them that we are taking all the required measures to implement the enacted laws diligently. The statutory authorities may conduct their own audits on organizations and when they audit our organization, we need to represent our organization with them and ensure that they award us a clean certificate without attracting any penalties. Statutory audits are usually unannounced and unscheduled audits aimed mainly at catching the organizations off-guard and see if the organizations are flouting any norms. Statutory audits are a very serious action and we need to ensure that the audits pass our organization without fail. Therefore, we need to maintain all records meticulously and be ready for the statutory audit almost at the drop of a hat. Normally, if we get one-hour notice for a statutory audit, we should consider ourselves as lucky!

In short, we have the onus of ensuring compliance to all applicable statutes in our organization, maintain all the records meticulously and diligently, conduct internal audits to ensure compliance and then represent the organization with the statutory authorities and keep them from penalizing our organization.

7.9 Grievance Resolution

Grievances are part of organizational culture. Grievance is the dissatisfaction of an employee or a group of employees about some perceived injustice or unfair treatment. Organizations overtly try to be fair in treatment of their employees. In the present day, organizations have put in place a code of conduct for all employees including avoidance of any kind of harassment to any employee. Still, grievances do crop up even though their rate of occurrence has come down significantly since the turn of the century. We can broadly classify grievances into three categories:

1. Grievances toward the organization
2. Grievances toward the bosses
3. Grievances toward the colleagues

Now let us look at each of these categories.

Grievances toward the organization – When the employee joins an organization, he/she understands the organization and joins it willingly. So, where is the room for a grievance for an employee against the organization? These grievances do crop up especially during organization wide changes including wage hikes, promotions and reorganizations. During the yearly salary hikes, employees compare the received hike with each other, even though we maintain that the individual hikes are confidential. The employee, who received lesser hike compared to a colleague in the same cadre with similar qualifications and experience becomes aggrieved. Whenever an organization wide pay hike is rolled out, we will certainly have a few grievances on our hand. Similarly, we also roll out promotions every year. The ones who were denied promotions, especially when they had put in the same number of years in the present layer as the one who received the promotion would be aggrieved. While pay hikes and promotions are yearly routines, reorganization is not a yearly matter. We implement an organization wide rejig on a need-basis. It may happen once in three years or more. Reorganizations can be minor in which a readjustment is implemented to take care of an exigency or reorganizations may be a major operation when consolidation and redesign of organizational structure could be implemented. The grievances due to a minor rejig may not result in many grievances like a major rejig which could give rise to many grievances. These will arise from those who perceive that they were not given important positions commensurate with their qualification, experience and track record or that their juniors were given unduly more prominence.

Some of these grievances could be genuine and the rest could be just imaginary. An aggrieved employee would not be able to perceive his/her own inadequacy or accept the rationale behind the perceived injustice. Whether the grievance, in reality, is justified or unjustified, the employee becomes disgruntled. Some employees may formally raise a grievance and some may not. Here is an important question that needs to be addressed: should we or should we not address the grievance that was not raised? There is comfort in "letting the sleeping dogs lie" but we really do not know if the employee who did not raise a grievance is satisfied or dissatisfied. Some employees do not raise grievances explicitly, especially, if the organizational culture does not encourage raising grievances. They just lie low and await an opportunity to take revenge on whosoever is perceived to be the cause of such injustice. We at the organizational level involved in the hike, promotions and rejigs are well aware of those who could feel dissatisfied. We should not only address the raised grievances but also address all those employees who are in the similar situation as the person who raised the grievance. We should not use the principle, "the crying baby gets the milk". When a formal grievance is filed and is received by us, it alerts us about the view point of one section of our employees and in the resolution, we need to treat all those employees as part of the grievance.

Grievances toward the bosses – grievances toward the bosses are also common but these are individual grievances. It needs to be treated on a case-by-case basis. Grievances with bosses is a common occurrence. Bosses assign work, measures the results, counsel and coach for improvement in performance, admonish in cases of tardiness or for diminished quality of the deliverables and a host of other interactions in the execution of work. Some of these transactions can go awry and leave the employee aggrieved. One occasion in which maximum grievances happen is the time of performance appraisals. For any serious disagreement between the employee and the boss, a grievance arises. But most employees prefer not to file a formal grievance for fear of being victimized in future. When the grievance becomes unbearable, people normally resign and leave the company. For employees to file formal grievances, we really need to put in extra efforts in handling the filed grievances. A tactful handling of the grievance can send a message to the employees that they would not be victimized for filing a grievance. We also need to keep monitoring the employee and the boss to ensure that the employee is not victimized in any manner by the boss.

It is common to show some amount of lenience toward the boss as that person is holding responsibility for results and therefore the boss ought to have some rights over the subordinate. Admonishing a superior in the presence of a subordinate diminishes the respect the superior enjoys from the subordinates. If the subordinates lose respect for the superior, they may not obey the reasonable instructions of the superior and it is impossible to conduct all superior-subordinate transactions in writing. But sometimes the bosses could cross the line between a gentle admonishment and an insult. What we need to do is to do justice and also make it explicitly so that the involved parties see the justice in it. We need to handle the grievance very tactfully but truthfully. When this kind of grievances come up, the ending should ensure that neither the boss nor the subordinate should remain aggrieved.

Grievances toward the colleagues – Grievance toward colleagues arises from harassment, bullying and conflict at the workplace. We may not be able to perceive it but bullying and harassment exist in the workplace. Conflict at workplace is well recognized and a lot of literature can be found on conflict management at workplace. Harassment, especially the sexual variety is now recognized and is included in the "zero tolerance" class of unacceptable actions at workplace. Now, most organizations have in place rules regarding sexual harassment and the punishments are severe. Bullying is not well recognized as reality in organizations but it does exist. Calling names, passing slurs, putting down and such other acts which happen more often at schools do also happen in adults in organizations. Sexual harassment is now being reported without any apprehensions. But raising a grievance for bullying and conflict is very rare even in these times. We need to recognize these acts because we will be losing good employees otherwise.

Grievance handling procedure – Why should we handle grievances at all? If we do not handle grievances what could happen? Well, the following are the consequences of unresolved grievances:

1. A grievance is raised when an employee is aggrieved. An aggrieved employee becomes disgruntled and demotivated given enough time. The damage that can be caused by a disgruntled employee is enormous.

2. If the grievance raised by an employee is not resolved, the employee may perceive the organization to be inimical to him/her and ultimately leaves the organization. An experienced and internalized employee leaving the organizations leaves a big dent in the organization. Sometimes the loss can be irreparable.

3. People have a tendency to settle their grievances on their own and in their own way. When the organization fails to resolve grievances, the grievances become disputes and the people may settle their grievances using informal means giving raise to Godfathers in the organization. This is a gravely undesirable development which can destroy the very fabric of organizational culture.

4. Unresolved grievances would cause the aggrieved party to commit covert acts of indiscipline and damage to organizational property or deliverables. The involved parties can engage in feuds which further damages the work environment which is not conducive to harmonious working and for producing desirable results for the organization.

Therefore, we should not leave any grievance unresolved or left to time to resolve those grievances. We should expeditiously and posthaste resolve the grievance and send a message that the organization is eager to resolve the grievances and move forward positively. Here comes the role of a grievance handling procedure. Here are the components of a robust grievance handling procedure.

1. An organizational entity is earmarked to receive the grievances, handle it through to resolution and follow through to ensure a positive ending to the grievance.
2. A robust grievance handling procedure that is periodically improved with the feedback received from the grievances handled in the organization.
3. Support and involvement of senior management in the form of funding and resources for the entire activity of grievance resolution.
4. An organizational environment that is conducive to raise grievances without generating any animosity and a resolution mechanism that does not leave any rancor for the parties involved. A grievance is treated more as a systemic issue rather than a personal issue.

While so, many organizations look at employee grievance as an affront to the management or a specific person and treat it as a personal issue. They feel that the employee is a troublemaker than an individual that has a genuine issue. Therefore, they prejudge the grievance and use authority and bullying to silence the employee than examining the issue thoroughly. In such organizations, once an order is issued, they do not like to rescind it or modify it. It is as if it is etched in stone. While it has been advocated right from Henry Fayol that employee tenure and loyalty are very important things for organizational prosperity, it is given lip service than real concern. This is a common pitfall. Organizations, especially its senior and top management has to be sensitive to the fact that grievances are a reality of modern organizational working and need to be resolved amicably. We can confidently aver that a proper grievance handling procedure that results in a win-win situation for the organization as well as the individual would go a long way in having harmonious relations in the organization leading to excellence in deliverables and sparkling results.

7.10 Working from Home

Working from home is a recent phenomenon which is being adopted in a big way since the turn of the century. When the work to be performed does not need a special high-cost equipment, or the team to assemble at one place or frequent interaction between colleagues is not essential or where there is no face-to-face interaction with the clients, it makes sense to work from home. It is the situation in software development industry, especially for the programming job. Any communication can happen on emails, phone calls and meetings can take place on video conferencing or teleconferencing. Now technology is available to conduct videoconferencing across oceans which can give the feeling that all participants are in the same meeting room. In videoconferencing we can make presentations as if we are making it in person. Therefore, it is not essential to sit in the same room or facility to perform the assigned work. Unless deliverables need to be handed over in person, one need not meet the others in person. In software development, documents and code artifacts can be emailed or uploaded to a common workspace designated for the team. Since there are advantages in saving money in the form of savings on rent, facilities maintenance, energy and communication costs and other related expenses, it is adopted in a big way across industries. The employees can be hired across the globe working on a common project sitting in different countries.

Now the question is, what is the role of employee relations in such a situation? While the involvement is greatly reduced, there still is a role, however miniscule it may be. Here are the aspects of employee relations function that need to be performed in an organization that has a significant number of employees working from home.

1. We need to administer their salaries.
2. We need to manage their career.
3. We need to resolve their grievances.
4. We need to ensure compliance to the applicable statutes.
5. We need to manage their LOA and holidays.
6. We need to administer their pay hikes.
7. We need to ensure that all organizational code of conduct is adhered to by all the employees, even if they are working from home.
8. We need to ensure that all those working from home receive all communications sent by the organization.
9. We need to ensure that those working from home participate in the organizational initiatives including those activities that are aimed at building bonhomie among employees such as company picnics, dinners and other programs.
10. We need to ensure that they participate in training programs and seminars as needed.
11. We need to ensure that they participate in any other organizational activities that may need employee participation like certification audits, internal audits and so on.

In this manner, we ensure that the employees, even while working from home, feel part of the organizational family to ensure high morale and to deliver quality products for our customers.

7.11 Software Usage for Employee Relations Function

Perhaps this is a question that should not be asked in these days. Registers, paper documents have bitten the dust except in rare circumstances. Computers have taken over the information processing and dissemination process. Therefore, excepting in very small organizations, all others use some software or the other. Irrespective of the size of the organization, the software needed to handle the employee relations function is not simple. This is because, the activities performed and the rules for performing such activities, the analysis needed and the reports to be generated are more or less the same. However, the software for handling employee relations function is not marketed in a stand-alone manner. It is usually bundled with the software that handles the entire HR function. It is usually referred to as HRMS (human resources management system) or something similar. Peoplesoft was the pioneer in this segment but others had also recognized the critical nature of this vital function as well as its market potential. We would not recommend any specific software that can be used by all. No one-size fits all. You need to evaluate the software and match with your needs and then make an intelligent decision. But we do advocate using some software or the other for the employee relations function. This allows you to carry out a variety of analyses on the data being generated.

8

Salary Administration

8.1 Introduction

The primary reason for people to work in an organization is to earn a livelihood for themselves and their families. Salary is the primary motivator of human beings in propelling them to work. Even when the individual is feeling like not working, salary is the compelling force that drags an individual to work and makes him/her put in the required effort using the best skill at his/her command to earn the salary. Conversely, the salary can be the worst demotivator if administered poorly. Therefore, it behooves on organizations to administer the salary meticulously and diligently on time and every time. Therefore, paying the due salary is one of the very important duties of the organization. In this chapter let us discuss the function of salary administration.

8.2 Organizational Framework in Salary Administration

Before we embark on the matter of salary administration, we need to discuss the organizational framework that needs to be put in place. The organizational framework as discussed in Chapter 2 of this book consists of policies and procedures that govern the organizational working. Individuals in the organization work within this framework performing their duties and achieving the expected results. When it comes to salary administration, the organizational framework consists of the following factors:

1. The specification of categories of employees, namely, employees on the regular rolls, temporary hires, consultants, trainees, and any other organizational specific employees that can be employed in the organization. For each of these categories, the pay ranges that can be paid need to be defined. Of course, they need to be revised periodically to keep those pay scales on par with those in the industry.

2. For all categories of employees, we need to define the number of levels from top to bottom. The bottom level would be the working level employees. That is, they carry out productive work and deliver results. Then the supervisory level that assigns work to the working level employees and closely supervises the work. Then we have the managerial level which manages the work of a single department. These people would plan, organize, staff, and coordinate to produce the desired results. While the supervisory level is responsible to ensure that work is carried adhering to the specifications, at the desired productivity and quality levels, managerial level is responsible for integrating the work to produce the deliverables, as well as to take overall responsibility for results. The next level is the senior management level comprising of those that head a profit center or a large department with independent charge responsible for financial results and competitiveness and continuation of the operation into the future. The top most layer is the top management which has the responsibility for the strategic management of the company.

3. In each of these levels, we need to define the number of layers that can be there in each level. Why layers in a level? People need some advancement to look forward to staying motivated besides salary. Basic salary guarantees penalty-avoidance level of performance. Hope of a future promotion and advancement of career would propel them to put in higher level of performance on the job. On an average, the employee would have about 35 years of service to advance his/her career. When we recruit resources, we recruit candidates that have the minimum qualification for the level. Suppose, if we recruit a person at the working level, how far can we advance his/her career in the 35 years that person has the career? Obviously, that individual cannot reach the top management position! A person can certainly raise to the next level with right approach and diligent performance. Exceptional candidates may rise two levels above the original level. Rising 3 levels or more is rather an extraordinary event. No one would like to stay at the same level or just one level above the original level for 35 years! Therefore, the organization needs to provide a career advancement option once every 3 or 4 years so the employee feels happy that his/her performance is being recognized. That is the reason for dividing each level into layers. Usually, we have three levels (junior, normal and senior) or four levels (trainee, junior, normal and senior) or a similar classification with suitable designations. For each of these layers, we need to define the minimum

salary that needs to be paid and the maximum salary that can be paid. This will help us to fit the person while recruiting him/her into the organization.

4. The periodicity and timelines for paying the salaries. Periodicity is the interval between two adjacent salary payments. They could be weekly, fortnightly and monthly. The timelines would be the first/last day of the week/fortnight, month or a specific date(s) on which the salaries would be paid to the employees. This specification would aid the employees in planning their expenditures and running their homes.

5. Then, we also need to define the components of the salary. The salary could comprise of a basic rate, then contributions for medical insurance, contributions to retirement plans such as the 401(k) plan, and some yearly components like the personal time-off, leave travel assistance, performance bonus, defined benefit pension plans and so on. These will enable the employees to save money for their future security.

6. We also need to define the special allowances which include allowances toward official travel, long duration working at inhospitable environments outside their normal location, working in hospitable but client locations, for working extra hours at their normal location as well as at client locations and so on. This would give predictability to the employees as to what they can expect from the organization for this kind of exigencies.

7. We need to define how we would assist our employees during times of distress like hospitalization with a serious ailment, natural calamities and disasters, accidents when on duty and so on. These will bring predictability to the employees what to expect from the organization in extraordinary circumstances.

8. Any organizational assistance to employees to acquire big ticket items including housing and so on so that the employees can be motivated further to put in tenure with their best efforts in the service of our organization.

9. Any other aspect of employee payments that are organization or industry specific like membership fee for professional associations like the IEEE and others.

Once we define all these aspects for employee payments, the organizational framework is setup. Of course, we keep revisiting this framework periodically to assess its impact and improve it based on the feedback and the benchmarking with comparable organizations in the industry.

8.3 Regular Salary Administration

Salary administration begins with fixing the salary of the employee at the time of recruitment. Of course, it would be handled by the recruitment wing of HR department following the guidelines set forth in the organizational framework for the salary administration. Employee salaries are negotiated with each employee at the time of his/her hiring. The fixing of the salary depends on several factors including the criticality of that employee in our organization, the skill set, the pedigree which includes the qualification, the educational institute from which the individual graduated, the previous organizations, the skill set and so on. In organizations, we have about ten salary bands with a minimum and a maximum salary for any specific level. When we make an offer to a prospective employee, we place that individual in a salary band in between the minimum and the maximum allowed for that band. Sometimes, when the candidate demands a higher salary that does not fit within the band and we cannot take that individual in the next higher level, we offer a special one-time cash payment as a joining bonus or stock option or something similar which could get the employee in without the necessity of changing the level. So, each employee in the organization may have a unique pay package which needs to be paid on time, adhering to the contracted salary. Another aspect is the confidentiality that needs to be maintained about the pay details. Salary is a sensitive subject for the employee as well as the organization. The employee may feel belittled if his/her salary is less than the colleague's. Other employees may feel jealous or aggrieved if they perceive any other peer to be drawing unduly higher salary. This is detrimental to both the organization as well as the individual. Therefore, keeping the salary details confidential, is a very important matter.

In some cases, the salary is paid by an hourly rate. The salary for the period would be computed using the formula:

Total salary = Hourly rate × Number of hours worked

In some cases, the employee would draw a monthly salary without any reference to the number of hours worked. Most managerial positions and some professional workers would come under this category. For this class of employees, it is usual to provide some allowances for payments for reimbursing the expenses that are allowed by the organizations. Here is a list of popular allowances allowed to employees in organizations:

1. In most organizations some scheme or the other would be in place to help the employee with the medical expenses for self and family. Some

organizations provide an insurance cover and pay full or part of the premium so that the employee and his/her family can have affordable health care.

2. For scientists and professional workers, membership fees to professional association would be reimbursed.

3. For employees, some organizations pay special pay to keep their knowledge up to date in their field. For those who must maintain their certifications or licenses such as Registered Financial Advisors, Accountants, Legal Counsel, PMP (Project Management Professional), CFSP (Certified Function Point Specialist) or similar certifications that mandate the professional to put in a certain number of PDUs (Professional Development Units) to retain that certification.

4. For certain employees, leased accommodation or reimbursement of house rent is provided.

5. Some organizations provide some sort of subsidy or reimbursement for vacations. While LOA needs to be planned and obtained by the employee, travel cost to a selected location for the annual vacation may be reimbursed in total or partially by the organization. This allowance is usually referred to as leave travel concession or something similar to it. This is usually allowed once a year or once in two years depending on the organization.

6. For certain employees, special allowance is paid to work in difficult working conditions. In mining industry, it is common to pay special allowance for those working inside mines. In software development industry, it is common to pay special allowance for those working at client sites during installation and commissioning of the software product. Similarly, it is also common to pay special allowance to those deputed overseas to work at client sites.

7. In inflationary economies, it is usual to pay DA (dearness allowance) to compensate for the increase in the cost of living. This allowance is reviewed half-yearly or quarterly based on the increase in the cost of living index for the region.

8. For certain positions which need to interact with customers or such other organizational guests, entertainment allowance or something of that sort would be reimbursed.

9. In some cases, a special pay is given to some employees. For example, a certain employee holds a higher office because that position falls vacant temporarily because the individual holding that position is absent for a long duration. The absence could be because the individual is on long

LOA due to medical conditions, a long tour abroad or taken a long leave to pursue education or passion or something like that. In such cases, the position would not be filled by another person but is assigned to someone junior to hold fort till the individual returns. Such people would be paid this allowance.

10. Due to exigencies of work, some people may be asked to work extra hours on a working day or work on a holiday. In such cases, some organizations pay extra money referred to as overtime pay which is paid at double the usual rate. In some organizations, the extra working is absorbed as an organizational necessity and no extra pay would be paid. In some organizations, in lieu of the extra work put in, a compensatory time-off would be allowed as a leave of absence.

11. There could be situation specific additional allowances paid to employees and the employee relations needs to ensure payment of such allowances.

While the finance department makes the actual payments, all these details need to be provided to finance department by the employee relations and also regularly update these details just before the salaries are processed for every pay period.

Besides other earnings, we have salary deductions to administer. In handling deductions, the employee relations need to not only deduct the money but also credit it to the intended agency and maintain meticulous records because some of the payments would be of extremely long durations which may span the tenure of employment that may last until the employee retires. We need to effect deductions in the salary for several causes which are enumerated below:

1. Of course, we do not pay for all the hours not worked in the month for those on hourly-rate but for the fixed salary people, we need to administer deduction for the unauthorized absence from duty. Usually, we adjust LOA for periods the employee is absent from office but if the employee used up all the paid leave allowed to that individual, we need to effect deduction for such periods of absence. Such unauthorized LOA, the organization depending on its policies may treat that LOA as paid LOA or LOA without pay or LOA to be adjusted against LOA that would accrue to the employee in the future. Employee relations department needs to handle LOA without authorization.

2. Most organizations have some plan for retirement savings like the 401(k) of the US. It may be called by different names in different countries, but

most countries provide such schemes some of which are supported by governments. The installment for such a scheme forms part of the salary deductions.

3. Some organizations provide a pay-linked pension scheme in which the organization matches the employee contribution and pays it to a selected insurance company to protect and administer the scheme. In some large organizations, the organization itself may administer the pension scheme. It is also a deduction from the payroll.

4. Employees would be provided with some sort of health insurance cover. Its cost may be totally born by the organization or the employee may have to bear a portion of that cost. In some cases, the organization may bear the entire cost for the employee and his/her immediate family and may allow the employee to include members of his/her extended family provided the employee bears the extra cost. This benefits the employee because the special rate allowed to the organization is charged to the additional members. The employee would save the difference in the premium rate charged to the private individuals which is higher than the bulk rates charged to organizations.

5. The organization may run some thrift schemes through which the employees can save money deducted straight from the salary and credited to the thrift account from which the employee can draw the required funds in times of need or even borrow! This results in payroll deduction.

6. Some organizations have special arrangements with other organizations marketing various products from consumables, groceries to big ticket items to provide special discounts on purchase to their employees guaranteeing installments checked off from the salary. It is another deduction for the payroll.

7. Some employees borrow money from banks and other financial institutions to buy or build a house for their family. In some cases, the organization aids the employee in paying installments deducting from the employee salary. Such arrangements encourage the lenders to give a small discount to the borrower because of the payment is not only guaranteed but also the periodicity is guaranteed. This is another deduction for the organization.

8. In some cases, the courts may order wage garnishments especially for various loan installments, child support payments or some other payments that need to be garnished from the salary and credited to the court or its authorized agency. This is yet another deduction for employee relations to handle.

9. Employees are paid various advances, the most common one being the advance paid to employees for meeting emergency expenses such as illness to family members or for recovering from the impact of a natural calamity like a hurricane. Such advances need to be recovered which are another deduction for the employee relations to handle.
10. If there are unions, the union membership fee needs to be checked off from the pay and credited to the union. That is another deduction to be handled by the employee relations.

The employee relations department needs to handle all these deductions efficiently. Just deducting the money from the salary is not the end of the story; it needs to be credited to the intended agency and records of those deductions and payments need to be maintained in a meticulous manner that need to be retrieved as needed. While the finance department handles the actual deductions and payments, they would only follow the instructions of the employee relations and all responsibility for all actions of deductions and payments rests with employee relations.

Paying the right salary after adding the eligible allowances and affecting the necessary deductions is one prime responsibility of employee relations and paying it on time is another responsibility as part of salary administration. Some organizations pay wages on a weekly basis. But, mostly, salaries are paid either on a fortnightly basis or on a monthly basis. Whatever is the adopted cycle for paying salaries, we need to ensure that salaries are paid on time. In some special cases, we may need to pay salary in advance of the pay date. For example, a resource comes to our organization from our branch office (or goes from our headquarters to our branch), we need to pay his/her salary before his departure date which need not coincide with our pay date. There can be other occasions like this when we need to pay salaries on dates other than our usual dates and we need to ensure that all salaries are paid on time and every time. Again, while the finance department makes the actual payment, employee relations needs to coordinate with all the concerned agencies namely, the resource, the supervising officer and the finance department.

It is common among organizations to defer a portion of the salary to year end under the umbrella of bonus. This pay component is also referred to as performance bonus, performance-linked incentive, profit sharing, sales commission, production bonus and so on. By whatever name it may be called in the organization, we need to note that it is the yearly component of the salary. Now, in some cases, the amount paid under this head is uniform as a percentage of the yearly earnings and in some cases, it may be linked with the performance of each individual. If the rate is uniform for all the employees, it becomes very easy to compute and pay the amount. While it makes the job of the finance department easier, the employee relations department must work hard to get all the stakeholders to agree on the percentage. It may involve bargaining with the unions, with the management and in extreme cases with statutory authorities too, if needed to get all of them to agree on the percentage. If each individual gets a different amount based on the performance, employee relations department needs to arrange for computing the performance of each employee and then get it accepted by each individual before making the payments. This can take time. Therefore, we need to take a couple months from the end of our financial year before we can effect payment under this head to the employees. Arranging the payment of the yearly component of salary is another task for the employee relations department.

Normally, salary is paid through bank in organizations except where the wages are paid every week in cash and that too when mandated by the law. The bank payment can be either by direct deposit into the account of the employee or by check. When all employees are at one location in a city, we use the direct deposit method. But if the employees are geographically distributed, we may send checks by mail to reach them on or before the due date of receiving salary. Does any organization pay salary using some sort of money order? It may seem strange, but they do, especially in software development organizations! When we depute our resources for installation and rolling out the software, train the end-users and hand over the system to the customers for short durations like 3–6 months, we cannot expect our employees to open bank accounts in those places to have access to a robust banking system especially when the resources are deputed overseas. In the US, the individuals need not have a bank account everywhere as they can use credit cards or debit cards and cash dispensing ATMs are easily accessible everywhere. But it is not true in all countries. When a resource is deployed overseas, it is possible that there may not be any access to a robust banking system comparable to that of the US. In such cases, we do send money through a money order using the postal delivery or an organization like the Western Union.

So, at every due date for payment of salary, we need to collate all these cases and forward them to the finance department sufficiently in advance so that they can pay the salaries on time and at the right place so that our resources are not put to any difficulty in receiving their salaries. While paying salaries on time at the right place in the right medium, may not get us any positive marks in terms of morale and motivation, but, if there is even a slight error, it could result in a serious backlash in terms of motivation, morale and confidence. Therefore, utmost care needs to be taken to effect salary payment, accurately, on time, and using the right mode of payment.

As we discussed in the previous paragraphs, our resources could be working at the headquarters, or branches or at client sites. When our resources work at places other than their usual place of working, the payments would be made from the location they work in. If a branch employee works at headquarters or a headquarters employee works at a branch or a branch employee working at a different branch, it is possible that the salary for that resource is paid locally. But the cost needs to be apportioned to the right project or cost center. The actual payment and apportioning the cost to the appropriate cost center or project would be carried out by the finance department. But we at employee relations need to intimate the details of the cost center to which the cost needs to be apportioned when we request the finance department to effect payment along with other details of the payout.

8.4 Salary Reviews

Sometime ago, the salary of an employee was considered only when the job description changed. The revision need not necessarily be upwards, it could be downward if the responsibility or the productivity declined. The salary was hiked only if the productivity increased or additional duties were assigned to the person. But with the change of times, now it has become customary to view salary review as a hike only! That is, the organizations must perforce increase their profits and hike the salary to retain their existing employees. Well, all the governments adopted inflationary economics and price rise is the regular matter not needing any discussion. It is taken for granted that prices rise, which raises the demand for higher salaries which will push up the prices further. It is a vicious spiral. All the same, this situation is here and would continue for the foreseeable future. Still, downward revision of salaries is carried out in extreme cases when the organization is passing through difficult financial times. But it is rare and when it happens, the salaries would be cut from top level percolating downward to tide over the situation and when the situation improves, the restoration begins at the bottom level and percolates upwards!

For salary reviews, we need to consider three important criteria, namely, the capacity to pay, the management philosophy and the need to maintain parity with the salaries being paid in similar organizations in the vicinity.

When we compare the salaries in our organization with those of others, we need to take the following into consideration.

The organizations need to be similar, especially the nature of the deliverables, the target markets, the size and the location.

If one organization is engaged in contract development and the other is a product development, they are not comparable. Contract development would be priced for every contract in a different manner depending on the competition at hand and the customer's capacity to pay and the need to win the contract. It follows the opportunity and cost-plus based pricing model. In this model, the profit in one project would be different from the profit of another project. It could be either more or less than the profit of the previous project. The product development organization uses the breakeven pricing model. If the marketing could sell more than the breakeven volume, the organization makes more profit and the profit increases directly in proportion with the volume of sales. Therefore, if we are a contract-development organization, we need to compare our salaries with those of another contract-development organization but not with a product-development organization.

The profits of the organization would be different if the target markets are different. Let us consider an organization in the US which has target markets in the US and another organization whose target market is in India. Obviously, the organization that has its target market in the US gets better profits than the one whose target market is in India. Therefore, the organizations with customers in the US has better capacity to pay than the other one. So, while comparing our salaries, we need to compare them with an organization that is not only carrying out similar work but also within the same target market.

The profits of the organization would also depend on the size of the organization. If an organization is employing 250 revenue-earning staff and the other is employing 2,500 revenue-earning staff, the overheads would not be similar for both. The smaller organizations need to employ specialist staff even though the workload does not justify the position. For example, finance head needs to be of the same qualifications and expertise in both the organizations with almost identical salary. While the cost of such specialists is spread over the earnings of 2,500 employees, it would be spread over only 250 employees in the case of smaller organization. So, while the absolute amount of overheads to be absorbed may be similar but the percentage would be much higher. Therefore, we need to compare our salaries with those organizations that are employing similar number of employees.

The location of the organizations is also important in this matter. For example, the organizations located in the Silicon Valley cities like the

San Jose and San Francisco need to pay much higher salaries owing to higher cost of living than those of say, Upstate New York towns like the Poughkeepsie or Rochester where the cost of living is much lower. So, when we are comparing salaries of our organization, we need to compare them with another organization that is located in our close vicinity in addition to its nature of deliverables, target markets and size.

Even when we select the right organization to compare our salaries with those paid in our organization, we need only to maintain a parity only. We need not pay more than them unless we wish to project our company as the best pay master. Parity indicates we can vary our salaries by a small percentage such as 10% from those of the comparing organizations. One important point to consider here is that all organizations in our vicinity carry out pay hikes almost at the same time, which normally takes place after the end of the financial year and the final accounts are approved. While we know what they did in the last year and their present salaries are, we cannot know what hikes they would be implementing this year. We need to depend on multiple sources for this information. Industry grapevine is one source. While grapevine is not reliable to a "t", we can use it as a base and work our salaries. Second is the percentage hikes they have been implementing in last two or three years. It gives us an indication of what they are going to do this year. Third is the professional associations in which we meet our counterparts from other organizations and we can mutually share the information. You may be surprised, but this is a very widely used source. We give some information and they give some other information. By discreetly talking to many of our co-professionals, we can get a wealth of information. Professionals share information because, everyone faces the same quandary about how much to hike the salary and nobody wants to be wrong!

Once we compare the salaries with those of the similar organizations, we need to consider the philosophy of our management. What has management philosophy got to do with yearly pay-hikes – you may wonder. Plenty, we would assert. Some managements want their organization to be the best pay master of all other organizations in the country/region; some managements want that they like to be less than the best but better than the worst and would like hover either near the top or above the worst; and some managements want to position their organization exactly in the middle of the continuum. There could be variants of these philosophies by being the best among those in the selected product; selected size; selected geography; and so on. We need to take this into consideration while working out the pay hikes for the coming period and present them to the management.

Finally, we need to consider the capacity to pay. If we have more than expected profits and less than expected expenses, that would be the ideal situation. But more often than not, we struggle with the reality which has no obligation to meet our expectations. We wish and try our best to obtain the best profits for the organization and wishes would not always come to fruition. Sometimes, our profits would be less than the planned profits. Our revenues may fall, or the costs may overshoot our expectations or unforeseen emergencies may affect our profits. If we have the capacity to pay, we would gladly pay with management approval and the buy-in of our resources, but if our capacity is hampered due to paucity of funds, what should we do? Here is what we need to do.

First ascertain, if the fall in capability is purely a temporary phenomenon which could be due to unexpected delays in deliveries or is due to a serious cause that could stretch our incapability over longer periods. If the shortfall is a temporary one, we can ignore it. But if it is going to be longer run, we need to tighten our belts.

We need to consider the importance of retaining our resources. Smaller organizations have a much greater need to retain the employees as replacing is not only costly but also seriously affects our ongoing projects and cause further losses or even customers. We may not be able to withstand the event of losing senior and critical employees. Larger organizations have much better resilience to absorb the loss of senior employees and they can fill the gap with other equally competent internal resources until a suitable replacement can be found. In case, we are from a smaller organization and cannot afford losing employees, we need to offer better pay hikes, even if we cannot really afford them. But if we are capable of absorbing the loss of employees, we can cut the pay hikes to an affordable limit.

If our profits overshoot our expectations and have plenty of funds to affect our pay hikes, we can go ahead and give higher pay hikes to our employees. One word of caution needs to be stated here: If we implement out-of-the world higher pay hikes, it could have two effects. One is, it can create an impression that the management will willingly share the profits with the employees leading to higher confidence in the management and higher morale. The second one is, we can create an impression in the minds of the employees that they are entitled to higher pay hikes and may expect such hikes every year, year after year, even in our lean times too. This could create an entitlement-mentality in our employees, which is not conducive to better working conditions. This is not to say that we should not give higher pay hikes in times of higher profits but to caution that such higher pay hikes need to be given carefully keeping the percentage pay hike to be at the normal level and giving some bonus in some form or the other.

In this manner, we need to consider all the three aspects into consideration before finalizing the salary hikes. Now we can go about determining the actual quantum of the hike to be given. To do so, we need two figures: one is the actual salary we have been paying presently and the second is the figures from the industry. Internal figures are available at our fingertips from our finance department taken straight from our computers. Where do we get the industry figures? Most common practicc is to conduct salary surveys to obtain the figures from the industry. We now have many consultancy organizations offering to conduct salary survey for a fee. We need to select the industry type and they will conduct the salary survey from those organizations and provide us a detailed report conforming to our specifications. The report can be a summary or a detailed one including the individual responses depending on our specifications. The charges would also vary with our specifications.

Alternately, we can also conduct a salary survey using our organizational infrastructure. In the earlier days, it was well-nigh impossible to conduct a salary survey on our own, because, to conduct a survey, we need to physically travel to the selected individuals and take down their responses. With the advent of internet and web-based surveys, the job of conducting surveys has become much simpler and easier. There are many consultancies offering email lists of various categories at very affordable prices and we can purchase the desired email lists and use them to get the responses. Of course, some of the email ids could be wrong but that would be about 10% and absorbable. We need to offer some incentive to the respondents to motivate them to respond and provide answers to our queries. Generally, the final report containing the survey results is offered as incentive to answer our survey. In some providers of web surveys, there would be a facility to mail the report to the respondents automatically. If the survey provider does not provide such a facility, we need to email the report to the respondents.

Most survey providers facilitate data analysis tools to analyze the responses provided by the respondents. If this facility is not available by the survey provider, then we can analyze the data ourselves. MS-Excel or a similar spreadsheet application is a useful but simple tool with which we can analyze the data. We can filter the data using multiple filters in spreadsheet application to short list the desired data to draw the right inferences. How to conduct the data analysis is beyond this book but it can be easily learnt from any good book on statistical inferences.

The advantage of conducting the survey ourselves is that we can have reliable data and subject it to any type of analysis desired by us. But we may not be able to generate the number of responses to get statistically significant number. But do we really need a very large number? Not really! For a category, three responses per organization from about five organizations would give us the required data. Having a higher number of responses would reduce possibility of error in analysis. The second advantage of conducting the survey using internal resources is that it costs much less than that of a professional survey. Third advantage is that the whole process is under our control. The only flip side is that the size of data obtained may be comparatively small and we do not have access to professional data analysts.

Once the survey is complete, we will have a fair idea of what the industry is paying to the employees and then we can fix the salaries for our employees taking into consideration the three criteria discussed in the previous section, namely, the capacity to pay, the management philosophy and the need to maintain parity with the salaries being paid in similar organizations in the vicinity.

While setting the quantum of hike, the following criteria need to be applied. While revenue is notional, costs are certain. Whether we earn the projected revenue or not, we need to pay salaries on time. Therefore, we need to have reserves or sources from where to raise funds if need be, to pay salaries on time.

When we increase the salary to a level, we cannot, under normal circumstances, decrease it in the coming years. Therefore, we need to set the hikes in such a manner, that we can sustain them in the coming years. If we increased the salaries by 10% this year, next year, the resources would expect the hike to be 10% minimum! Therefore, if we have bumper profits, we need to curb the tendency to increase salaries by a hefty percentage but give a lumpsum (or some other equivalent) in lieu of the hefty increase. The resources would get the same amount of money, but the lumpsum is not the same entitlements as the salary is.

We should also keep in mind that the resources would expect their salary to be on par with their peers in the industry. We, in our eagerness to keep the entitlement at a minimum and supplement the salary with big lumpsum payments, it may prove counterproductive. We should maintain a careful balance between employee satisfaction and the quantum of entitlement.

When fixing the quantum of hike, the finance department would be involved in the process as they would be able to tell us the capacity of the organization to pay the extra amount. They would apportion the earnings of the previous year to various expense heads and show us the amount that can be allocated for the purpose of salaries. They would also consider if extra funds need to be arranged and the cost of those funds which include the interest on the loan taken and other charges associated with borrowing money. Taking all this into consideration and with the guidance from the senior management of the organization, we fix the salary hikes. While other agencies are involved in fixing the quantum of the salary hikes, the administration, of those hikes, rests with the employee relations department. In fact, the role of the employee relations department is that of coordinating the entire effort, rolling it out and administering the pay hikes.

Once the amount, available for pay hikes, is determined with the help of the finance department and the senior management, we compute it as a percentage of the total money spent on salaries of the previous year. Then we use this percentage to set aside the money for each level. While the percentage is the same for all levels, the absolute sum would be different for different levels. Suppose, we spent $1 million last year, and the money allocated for pay hikes by senior management and the finance department is $100,000, then the pay hike can be 10% of the existing pay packets.

While fixing the quantum for each individual employee, we take into consideration the performance appraisal of the employees. We classify the performance of the individual employees into three or four classes. Then the money earmarked for the level is now classified into three or four percentages. Each class of performance is allocated a specific percentage raise over the existing pay. We may assign 6% lowest performance rating, 8% for lowest-but-one performance rating, 10% for top-but-one performance rating and 12% for the top-rated employees. An example would better illustrate the method:

1. Let us assume that the total amount of money spent on salaries for software engineers was $100,000.
2. Percentage increase for pay hike approved was 10%, that is $10,000 for the next year.
3. Let us assume the number of employees in the software engineer category is 10. The ratings are:
   (a) Top rated software engineers are – 2
   (b) Top-but-one rated software engineers are – 2
   (c) Lowest-but-one rated employees are – 5
   (d) Lowest rated employee is – 1

4. Now let us compute the planned expenditure using the percentages detailed above.

(a) Top rated software engineers are $-2 \times 12\% \times 10{,}000 = \$2{,}400$

(b) Top-but-one rated software engineers are $-2 \times 10\% \times 10{,}000 = \$2{,}000$

(c) Lowest-but-one rated employees are $-5 \times 8\% \times 10{,}000 = \$4{,}000$

(d) Lowest rated employee is $-1 \times 6\% \times 10{,}000 = \$600$

(e) Total expenditure on pay hikes in this manner is \$9,200

(f) This leaves us a balance of \$800. We can adjust the percentages of the employees and spend it or use it for some other kind of perk or a gift. The advantage in giving a gift is that it is tax deductible for the company and not taxable for the employees. A gift can be the same for all employees and it motivates employees much more than a pay hike of a few more Dollars.

5. Normally, the differential in pay hikes that the top-rated employees get is twice the lowest rated employees, but different organizations practice different apportionment rules.

In this manner, we work out the pay hikes for each of the employees. While the percentage hike is same, but the individual salaries are different, the individual employees get different sums of pay hike in absolute terms.

Whether we should give same percentage pay hike or same absolute amount for the employees in the same class of performance rating is the discretion of the management and both are used across the industry. Each can motivate or demotivate the employees based on the way we present the pay hikes to the employees.

Normally, the differential in pay hikes that the top-rated employee gets is twice that of the lowest rated employees, but different organizations practice different apportionment and differentiation rules. This decision is usually taken every year by the management. If in the perception of the management, the top-rated employees need to be encouraged, they may give a little more pay hike. Some managements feel that lower performance rated employees ought to get a message that their performance has been noted by the senior management and give a much lesser pay hike as a means of conveying that message so that those employees perk up and improve their performance in the next cycle.

It is common that immediately upon rolling out the pay hikes, grievances do crop up. Employees feel that they were given less hike than they deserve. They do so by comparing their hikes with the hikes received by their peers.

While we advocate confidentiality and admonish any leaks if caught, it is normal that the employees share the hikes they received. Those that feel that their hike is justified or is more than justified, they simply keep quiet. Those that feel that injustice is done to them, raise their voice. It is the employee relations department that needs to handle the grievances and resolve them satisfactorily. If we implement the process diligently, then the hikes received by the employee are just, in most cases. Then we need to convince the affected employee about the fairness of the hike. This is part of counselling which will be detailed in Chapter 9. Executives working in employee relations department need to have reasonable ability in conducting counseling sessions. In some, rare cases, mistakes do happen, and some employees receive less than they really deserve. In such cases, it is better to correct the mistake, but only after investigating the case and ensuring that injustice was indeed done. Recognizing such instances, we need to fix a second date for rolling out a second installment of pay hikes to correct the mistakes committed in the first round. Normally, organizations do this after three months of rolling out the pay hikes.

Mistakes do happen on both the sides, that is giving less than deserved hikes as well as more than deserved hikes. Mistakes of this kind are pointed out by those that perceive that their hike was less. What should we do in such circumstances? What we need to do is to investigate the issue first. Once we realize that the employee was given excessive pay hike, we can implement these actions:

1. First, we need to counsel the resource about the excess hike. Otherwise, we will have a deeply disgruntled employee in our organization.
2. Reverse the declared pay hike and give a proper hike.
3. Alternatively, we can adjust this excess hike in the next cycle. Counsel the employee to expect a lesser hike in the next cycle.
4. Do nothing. Allow the employee to enjoy the hike. This may motivate the employee to better performance to justify the hike. But it may also send a message to the rest of the employees that the hikes are administered without due diligence.

What do we do about the employees who received less hike than they truly deserved? On this aspect, there are no two opinions. We need to revise the hike to match the performance. Normally, most organizations set a date for pay hikes and another date to examine the grievances and to correct the discrepancies. It is on this second date that we roll out the corrections in pay hikes.

Once the pay hikes are decided, we need to ensure that the employees would be satisfied or, at a minimum, they would not be dissatisfied. Can we really satisfy the employees with the pay hikes any time with any amount of increase? Let us say, not normally. Unless we have an unexpected bounty of revenues leaving enormous amount of funds at our disposal, we will not be able to give 25% pay hikes. Pay hikes, in normal conditions, would range between 3% to 10%. As long as our hikes are in this range, employees would not be dissatisfied. Pay hikes need to come from the profits earned. The shareholders, the governments and the employees all have a claim on these profits. Shareholders would want a higher percentage of dividends and the governments want a higher amount of tax amount as the tax percentage cannot be changed with the amount of profit. In some organizations, there may be lenders who lent us money either for our capital needs or working capital needs. We need to pay back the interest on the loans and a portion of the principal to reduce the burden of interest in the future. Then we need to take money from profits to put in reserves to ensure future financial safety as well as to procure capital equipment for replacing the obsolete equipment in usage. It is the same profit from which we need to draw money for improving facilities for the employees at work.

We need to take all these aspects into consideration while implementing the pay hikes. When we get higher profits, employees want a higher pay hike, shareholders want a higher dividend and governments want higher tax monies. We need to satisfy all. Governments and shareholders can audit our books using their authority, but our employees cannot. But employees can express their resentment in so many ways in day-to-day working. So, except in rare conditions, we cannot delight the employees through the pay hikes. Then how can we obtain the buy-in of the employees for the pay hikes we are going to roll out?

The answer is an honest communication with the employees about the gross profit earned before interest and taxes and then the net profit. Then we ought to give them the apportionment of funds in a transparent manner so that the employees can appreciate the sincerity in our efforts to do justice in the circumstances. Some organizations utilize the organizational grapevine for this purpose. We need to acknowledge that grapevine does work well to produce the desired results but, in our humble opinion it is not the right approach. Once the management uses grapevine, all others begin using the same methodology and rumors become the main mode of communication in the organization which is not conducive to honest interchange of information. This would be detrimental to the organization in the long run. We can put this in an organizational HR bulletin board, if we use one. If not, we can send an email to all our employees from either the HR department or from the top management. This communication need not wait until the pay hikes are rolled out. We can do it as part of internal declaration of past year's performance by the management.

By doing so, we can obtain the buy-in of the employees and eliminate dissatisfaction in all but the most skeptical negative employees. The employees will understand the overall situation and appreciate the sincerity of the management's efforts to do what is possibly the best.

8.5 Implementing the Pay Hikes

Implementing the pay hikes includes the following steps:

1. The prerequisite to begin the process is the declaration of financial results which were audited by the organizational auditors. This is carried out by the top management with the assistance of the finance department.
2. Preparation of the funds flow statement for the next year's operations by the finance department and its approval by the management. This statement tells the amount available for pay hikes.
3. Working out the pay hikes using the selected methodology for all classes of employees and finalizing them.
4. Obtain the preliminary approval of the top management.
5. Obtain the buy-in of the employees for the proposed pay hikes.
6. Roll out the pay hikes on the selected date. There will be a time lag between the end of financial year and rolling out of pay hikes which is usually 3 months. If the financial year ends on 31st December every year, then the pay hikes would be rolled on 1st of April.
7. Receive grievances raised by employees, if any, on the anomalies in the pay hikes. Investigate all of them and resolve them amicably. Consolidate and prepare a statement of all the resolutions that need to be implemented. Obtain the approval of the top management and pass on to the finance department for implementation on the specified date.
8. Roll out the final installment of pay hikes on the appointed day. We normally take three months from the date on which the first installment of pay hikes was rolled out for rolling out the second and final installment of pay hikes. If we rolled out the pay hikes on 1st April, then the final installment of pay hikes would be rolled out on 1st July. This closes the cycle of pay hikes for the year.

8.6 Terminal Benefits

The final aspect in salary administration is the administration of terminal benefits. Terminal benefits are those payments that accrue to an employee or his/her family when the employee exits the organization. An employee exists the organization due to resignations, termination of employment, death or retirement. Let us discuss the terminal benefits that accrue to the employee in these cases.

Resignation – An employee may resign from the services of our organization for a variety of reasons. When the employee resigns and is relieved, he becomes due for the following payments and deductions:

1. Salary for the last pay period of working in the organization which may be a week, or fortnight or a month.
2. Any yearly payments which may be paid on a pro rata basis. It would not be applicable for all companies.
3. Any company savings plans that become payable only when the employee leaves the organization.
4. If the resignation arises out of an injury caused while on duty, then the compensation for such injuries. Injuries can be caused by a variety of reasons including fires, electrical shocks, any accidents and so on.
5. Then there may be some deductions that need to be deducted from the terminal benefits due to the employee. They can be:

    (i) Any outstanding advance given to the employee for which account needs to be settled by the employee.
    (ii) Any loan given to the employee including housing loans, vehicle loans, healthcare loan and so on.
    (iii) Deductions for the final installments for the long-term savings plans like the 401 K plan.
    (iv) Deductions for company provided facilities like leased accommodation.
    (v) If the employee did not give adequate notice before leaving the organization, then we need to deduct money from the terminal benefits in lieu of the shortfall in the notice period as set forth in the employment contract.
    (vi) Any other situation specific deductions that may need to be deducted for the employee.

In this manner, we compute the terminal benefits that accrue to the employee when he/she resigns and leaves our organization. Then we communicate the same to the finance department for effecting the actual payment.

Termination – The organization may terminate the services of an employee due to various reasons and send the employee out of the organization. In such cases too, we need to effect terminal payments and deductions as discussed above in the case of resignations. When an organization terminates the services of an employee, the release would be immediate. That means, the organization did not give notice of termination in advance of termination as specified in the contract of employment. In such cases, we need to pay money in lieu of the shortfall in the notice period we need to give to the employee.

Death – Rare and a sad event it might be, death does occur while an employee is still in service of our organization. In such cases, the employee would not be present to collect his/her terminal benefits. Worse, the family members who collect these benefits would not even know what benefits accrue to the deceased employee. We need to compute the benefits and deductions that accrue to the employee and arrange for their payments. In addition to the benefits detailed in the previous two paragraphs, we may need to pay insurance settlements, and any other company specific benefits allowed for employees deceased while in service. Sometime, if the death occurred on our premises due to an accident, we may need to pay extra money to the next of the kin of the deceased employee. We may not be able to pay all the due benefits in one installment to the family. Insurance claims and accident policies etc. may take time to claim. We need to track all these and ensure that all installments are paid to the next of the kin and arrange for paying them. Sometimes, there may be more than one claimant for the benefits of the deceased employee. In such cases, we need to request the claimants to establish their inheritance through legal channels and produce an uncontestable claim to claim and receive the benefits due to the deceased employee.

Retirement – Retirement is akin to resignation with some differences. When a person retires, some companies provide for pension plans in addition to other benefits. Retiring employees are generally treated in a special manner as they established their loyalty to the organization. There may be multiple benefits that accrue to the employee in addition to the benefits that accrue to the resigned employee. They will be specific to the company. We need to compute all such benefits, collect all the insurance claims and decide to pay the benefits to the retiring employee. Again, we may not be able to make all the payments in one installment. If so, we need to follow up with all the concerned agencies until all the dues are collected and paid to the retired employee.

An employee still in service, can follow up with us and pester us till all the money due to him/her is received. But an employee that left our organization would not be within our organization to follow up on his/her claims. So, we need to be diligent in settling the claims of those employees that left our organization for any reason. We need to be especially empathetic in the cases where the departure is due to unfortunate cause of death.

The function of salary administration is very important for the organization. While the recruitment activity gets the best resources our organization deserves, it is the function of employee relations that retains them and nurtures them to be the best resources in the industry. There are other functions of motivation and morale that are also part of this process, but we will discuss them in the next chapter.

9

Motivation and Morale Management

9.1 Introduction

What motivates human beings and what keeps their morale at a high level? This is an age-old question. When we examine the history, especially the wars, we discover some very interesting details about motivation. A lot of research was conducted into morale and motivation and quite a few theories were expounded and the last word is yet to be pronounced on this subject. If we go back in time, in fact, too long back for historians and researchers to dredge up, we see wine, women and wealth as the main reasons for wars in the ancient times. The 20th century wars discredited wine and women as causes of wars. With democracy taking root in most developed countries, it is difficult to go to a war for the sake of a woman or in the stupor caused by excessive drinking of wine or another, perhaps more potent, alcoholic drink. Only monarchs or dictators were capable of waging wars on those reasons and their number steadily depleted to alarming levels. That leaves wealth. In ancient times, when a spade was called a spade and not as an excavating tool, they went to war for looting another tribe, state or kingdom. Now, we camouflage wealth as "natural resources" to refer to oil and mineral resources needed by us.

OK, let us for the moment, ignore the king/emperor/state going to war, and let us take a look at the pawns of war. What motivated the soldiers to go into an endeavor from which the fellow may never return alive and yet, what kept the morale high? In today's world, most people refuse to join army unless compulsorily drafted. Of course, there are exception in highly populous countries which have too many people and too few jobs. There, they see army as a career with assured income besides lifelong benefits including a decent pension. In the ancient times of Julius Caesar and Alexander, the soldiers were paid a pittance, if at all, but they had many willing soldiers to fight their wars. How could they get so highly motivated fierce soldiers at such a low cost? If we carefully sift the facts, the following appear to be the main reasons:

1. For invading armies:

    (a) The soldiers joined to have an opportunity to exhibit their masculinity in the form of valor and the number of enemy soldiers killed. This exhibition of masculinity won them many damsels.

    (b) They were usually promised with unlimited possibility to loot, arousing their greed.

    (c) They were also promised unlimited opportunity to commit rape, arousing their lust.

    (d) They were also promised unlimited opportunity to capture and bring back both male and female slaves for their service to use as they like arousing their need for luxury.

    (e) An opportunity to serve their own God and destroy the God of the enemies.

2. The defending armies were motivated:

    (a) To save their way of life and the threat to their wealth and possessions from the invading armies.

    (b) To protect their families and the threat to their women folk and families from the invading savages.

    (c) To fight for their God invoking their sacred duty and for luxury in after life.

3. To both the armies, their priests assured that a special place will be reserved in heaven as a favorite of their God with unlimited supply of beautiful and dutiful, drug and disease free, heavenly whores for their enjoyment along with heavenly abode, clothes, food and drink.

If we analyze and draw inferences, we see the reasons of greed, lust, threat and a priestly promise of a wonderful after life unavailable on earth. Now, we cannot seduce people to wage war with those enticements as people are not naïve anymore to believe that kind of stuff. The priests are a discredited lot and not many strongly believe in a wonderful after life. With international bodies like the UNO and intervention by superpowers either to prevent wars or to penalize the aggressor nation, the threats receded considerably. With looting and raping becoming punishable even in wars, that incentive lost its attractiveness. Females also joined the army and this has nullified the lure of rape after the war. Now as we noted earlier in this section, people join the army for a career or forced by the governments, not otherwise! Once in war, the soldiers, either in ancient times or the present times, have no other alternative but to keep the morale up and cooperate with each other as their lives are at stake!

Morale is the level of motivation in the team or the organization as a whole. *Motivation is an influence internal to an individual that elevates his/her performance to higher levels.* It is such an influence that makes one shed life if necessary. While motivation is internal to a person, it can be externally induced. Personal needs/wants motivate people into action to reach a goal. Some needs are physiological such as hunger, thirst and need for sex. Some needs are psychological. To fulfill those needs, people perform all the necessary actions. Deprivation of these needs elevates the motivation of people to unheard of levels. Alternatively, creating the feeling in a person that a certain need is deprived can also elevate the motivation of a person to greater heights.

When we come to business organizations, we have only one reason for people to join our organization to do some work assigned to them and that is to earn a livelihood for them and their families. Those natural motivators (fear and violence) which were used for centuries to keep the work force motivated simply became illegal. Until the end of the 19th century, the overseers whipped/slapped/beat the erring workers with impunity. Lightest punishment was firing from the job and it is no more a threat, provided the individual can get another job. Even in the 20th century, the threat of losing a job kept the workforce working as instructed. Come the 21st century, this motivator, the fear, was also lost to us to motivate the employees. That left us only one real motivator and that is greed. Hope of earning money in moderate amounts motivates people to put in a penalty avoidance level performance on the job. Only greed can motivate people to exert themselves beyond the penalty avoidance level of performance. But money is always in limited supply and we cannot dispense it as we like because there are competitors to keep our costs down and unions and labor laws which enforce fairness in rewards and compensation.

Now with all this background, let us now face the question, "how do we motivate the employees and keep a high morale?" That is the topic of this chapter. First, of the plethora of theories on motivation, let us discuss some very important theories on motivation and then analyze them and construct our own technique. Let us now proceed.

9.2 The Hawthorne Experiment

In the field of motivation and morale, the Hawthorne experiment was a significant milestone and it is very important to learn about it before we proceed further.

Hawthorne experiment was conducted by Professor George Elton Mayo of Harvard Business School and his team at the Hawthorne Plant of Western Electric company in Chicago. Actually, a series of experiments were conducted there between 1927 and 1932. The results of these experiments astounded everyone. It showed that the productivity (the rate of production) does not entirely depend on the wages and the physical working conditions alone. It further showed that the productivity depends on the perception of employees on how they were treated. It showed that the intangible emotional factors have as much or even more influence

Elton Mayo

than the tangible physical and logical working conditions. The management initially put together the team and Professor Elton Mayo to investigate the effect of lighting on the productivity. While the team conducted four experiments, two are of great significance to us. The first experiment was to examine the effect of the level of illumination on the productivity. The researchers isolated a team of six workers to conduct the experiment. After observing the workers working at normal level of illumination and they gradually increased it to the level of sunlight. Then they brought the illumination level gradually to that of moon light. Surprisingly, in both the cases, the productivity increased! Then a similar experiment was conducted on a second batch of workers. But in this case, the results, although were astounding, the productivity did not show any appreciable change either upward or downward. While our intention in discussing this experiment was not to give the entire details, we wish to bring out one fact that gave rise to what is known as "Hawthorne Effect". Hawthorne effect (also referred to as the "Observer Effect") refers to the tendency of some people to increase their effort when they are aware that they are being observed. The experiment also proved that the motivation (as manifested in productivity) did not depend on the physical (illumination) conditions but it has something to do with emotional factors.

The real significance of the Hawthorne experiment was to bring out the fact that people are capable of raising their productivity irrespective of the physical conditions. This led to a host of research and theories into the behavioral aspects of the work force. During the interviews conducted as part of the Hawthorne experiment, the importance of allowing a say in the matters, the style of supervision, freedom in the way work is performed, and a feeling that the work being performed is important have been articulated as some of the factors that impacted their productivity by the participants.

Gradually, the effect of the Hawthorne experiment seeped into the organizations and the principle of employee participation in decision making in matters that concern them has been implemented. This discussion on the Hawthorne experiment, conducted nearly a hundred years ago is to enlighten the readers about the importance of the behavioral aspects of the work place in motivating employees to higher performance.

9.3 Two-factor Theory

Professor Frederick Herzberg was a professor at the Case Western Reserve University and at the University of Utah. In 1959, he propounded Motivator-Hygiene theory, which is more popularly referred to as the Two-Factor Theory or the Dual-Factor Theory. According to Professor Herzberg, people are influenced by two sets of factors, namely the "hygiene" factors and the "motivators". The presence of hygiene factors does not cause the motivation to increase but the lack of them would lead to decrease in motivation levels. The pres-  ence of hygiene factors ensures the penalty avoidance level of performance. Motivators, on the other hand, would increase motivation by their presence, but their absence would not decrease the level of motivation in the employees. Presence of motivators ensure better than the base level of performance. The hygiene factors according to Professor Herzberg are:

1. Job security: predictability of retaining or losing a job based on a defined set of criteria
2. Work conditions: in which the employee can work safely and securely

3. Compensation: maintaining wage parity with comparable employees in the same or similar organization
4. Organizational policies and procedures
5. Type of supervision
6. Interpersonal relations

The motivators according to Professor Herzberg are:

1. Challenging and satisfying work
2. Recognition for one's achievements
3. Opportunity to do something meaningful
4. Involvement in decision making in the matters concerning the employee
5. Opportunities for personal growth

The four combinations of hygiene and motivators are:

1. High on hygiene factors and high on motivators: this is the ideal situation in which there are no demotivators and the motivators are present. The employees would have the highest motivation levels. They will fight to the end of the world for their company. They will produce the best deliverables.
2. High on hygiene factors but low on motivators: The employees are not dissatisfied but are not highly motivated. This produces an average level of performance which we can refer to as penalty avoidance level of working.
3. Low on hygiene factors but high on motivators: Employees are highly motivated but their performance is average because they are dissatisfied. The job gets done but that is that.
4. Low on hygiene factors and low on motivators too: This situation is not tenable. The company ought to be lucky to survive. The employees are deeply dissatisfied. They will jump ship at the first opportunity they get. They would not be able to finish the work leave alone building great deliverables.

Well, the world is not digital in nature; it is analog in nature. There can be many levels between high and low levels of the presence of the factors. How will motivation be, if the situation is moderate on hygiene factors and moderate on motivators? This, moderate on both hygiene and motivators, is the real-life scenario. There is no organization in the world that can provide "the best" in hygiene factors and there can be no great supervisors that can provide the best motivators! That is, the situation, hygiene and motivators being moderate, in which most of us work and in such cases, the motivation would be fluctuating. This theory was well received by the captains of the industry and the presence of hygiene factors is now more less the norm. But, have the employee complaints arising out of hygiene factors died down? Well mostly, but not completely. The human tendency is to expect the same conditions as the better places of work than the worse places. Complaints are muted but heard that they (hygiene factors) could be better! In the present day, because of the recognition of the importance of the hygiene factors by the managements, the focus is on motivators and we must acknowledge that managements are making sincere efforts in training their people to provide motivators to their employees at all levels.

The employees all over the world need to thank Professor Frederick Herzberg to bring about the change in the attitude of the managements and nudge them to provide better working facilities including pay and perks as well as to provide motivators for improving the motivation of the employees.

9.4 Hierarchy of Needs by Professor Abraham Maslow

Professor Abraham Maslow was an American psychologist. He classified human needs into five classes depicted in a triangle, which is depicted in Figure 9.1. At the bottom of the triangle are the physiological needs, which include breathing air, food, water to drink, sleeping for rest, excretion to flush out body waste and sex. Human beings spend maximum effort on fulfilling these needs. One level above physiological level are the safety needs, which include security of the body, employment, property, family and health. Human beings spend less effort on these needs than on the physiological needs but more than the other needs. The middle level needs are love/belonging needs, which include friendship, family and sexual intimacy. The top but one level needs are esteem needs which include confidence, achievement, receiving respect from others, and gaining self-esteem. The top-level needs are self-actualization needs. Self-actualization is understood as achieving the full potential of the individual. Self-actualization is becoming whatever a person is meant to be. Abraham Maslow wished that everyone self-actualizes himself/herself.

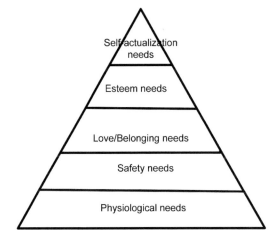

Figure 9.1   Abraham Maslow's Hierarchy of human needs.

The interpretation of this hierarchy of needs is that an individual would be motivated to exert himself/herself to fulfill these needs. Physiological needs take precedence over every other need and unless these needs are fulfilled, the individual would not bother about the next higher level of needs. Once the physiological needs are satisfied, the individual would begin yearning for the fulfillment of the next higher level of needs and would be motivated to work for the next level of needs that is safety needs. Safety needs are fulfilled when there is secure employment or work that can assure the individual of continuing satisfaction of physiological needs. Once these safety needs are satisfied, the individual would only be motivated by love/belonging needs. Love/belonging needs can be fulfilled by having a steady relationship or marriage and having a stable family and relationships. Esteem needs would be fulfilled by getting recognition in the society in general and by getting a rank at the place of work. It is basically fulfilled by getting respect from others. When that happens, the self-confidence of the individual grows and then the esteem need is fulfilled. Self-actualization is not that easy to be satisfied. Mere fulfillment of all needs till the level of esteem needs would not make the individual to yearn for the fulfillment of self-actualization needs. Most people would not realize that they are at the self-actualization level at all! In fact, a person does not look for anything from external environment when he/she is at the stage of self-actualization. Self-actualization is becoming what one can, ultimately be. A person at that stage would be looking at greater good than something for himself/herself.

As you can see from Figure 9.1, physiological needs are at the bottom of the triangle and self-actualization needs are at the top of the triangle. The amount of effort and time the human beings are willing to exert is highest at the bottom on physiological needs, which progressively comes down as we traverse the triangle upwards and least at the top that is the self-actualization. If a person is deprived on any of the physiological needs, he/she would be willing to do practically anything to fulfill his/her physiological needs. But if a person is deprived on self-actualization needs, the individual may not even notice such absence! That is why most business organizations keep the second level needs of safety in uncertainty. That is, while the job is secure with the employee on the regular rolls of the company; there is a clause in the employment contract that the employee can be terminated with some notice at the discretion of the management. This measure is to give the employee safety with a veiled threat. This really works well in most cases.

The significance of Abraham Maslow's work is that a person is motivated to put in extra efforts to grow from the existing level of needs to the next level of needs. This can be said to be the harbinger of career advancement and layering in the organizational levels discussed in the earlier chapters. But one question, how do we motivate our executives with stable families whose needs till the esteem needs are fulfilled? We need to look for something else. To some extent, the next one, the expectancy theory gives us a pointer to the answer to this question.

9.5 Vroom's Expectancy Theory

Professor Victor Vroom developed this theory for on-the-job motivation. In a nutshell, Professor Vroom summarized it nicely, thus, "*Intensity of work effort depends on the perception that an individual's effort will result in a desired outcome*". This is very important from the standpoint of motivating employees for higher performance, especially those employees whose esteem

Victor Vroom

needs are fulfilled. The outcome expected from the standpoint of the employee is on two fronts:

1. Official goals
2. Personal goals

Official goals are the targets set for the assignment on which the individual has been working and personal goals are the targets the employee sets for himself/herself. While there would be official goals to reach for every assignment the individual carries out, personal goals would not be there for every assignment completed by the individual. The personal goals can be two-fold again. One will be the satisfaction derived from being able to complete the assignment successfully. When we execute a challenging assignment that has high visibility in the organization, completing it successfully and basking in the limelight will be a reward in itself and people look forward to such recognition. The second would be to receive some sort of award for the exemplary success achieved by the individual. This recognition may or may not be in the financial form. It can even be a mention in the team meeting or at a higher-level meeting or an email sent to important functionaries in the organization about the success achieved by the individual. When the assignment is of the usual type without an opportunity to exhibit a spectacular success, the individual may not have special expectations. But the individual would certainly have special expectations if this normal assignment is the culmination of a series of successful normal assignments, and then there could be expectations, especially at the time of performance appraisal.

The general interpretation of this theory is that not meeting the expectations would generate dissatisfaction but meeting those expectations would not generate additional motivation to put in better efforts. This theory is put to use in emphasizing the need for setting the expectations right. If we set some expectations and then not meet those expectations, it would certainly lead to demotivation. That is the reason for the emphasis on setting right expectations.

9.6 Carrot-and-stick Theory

This theory is a much-discredited theory in motivation and morale subject. Generally, fear, represented by the "stick" in this theory is no more a motivating factor. Besides, there are rules and regulations against using a real stick in the organizations. Removing the stick from this theory retains only the carrot! That is giving positive rewards is the only option left today in this theory on motivation in organizations.

Come to think of it, the expectancy theory of Vroom, discussed above, sounds similar to carrot and stick theory! The employee would have some expectations from the efforts put in. Suppose, the employee did something that is against the regulations or standards or potentially damaging to the organization, what expectations would the employee have? Before jumping to answer this question, just think, what does a child do when he commits a mistake? The child would certainly expect some sort of punishment and would show it on his/her face! But adults would have the capability to camouflage their emotions and would not show this expectation on their faces for all of us to see. When the expectation of some sort of negative reward is not met, what would be the conclusion of the erring employee – one of relief and a confidence that the organization would allow transgressors to go scot free? This is detrimental to the general morale of the organization as the other employees also get that very same message that erring is OK. Therefore, carrot, positive rewards, for positive performance and negative rewards for negative performance need to be administered in the organization but in a careful, tactful and appropriate manner.

9.7 Stimulus-response Theory

This theory is credited to Ivan Pavlov, B.F. Skinner and others. This theory is based on the assumption that there will be a response to every stimulus. This can easily be checked in animal behavior. You give a cookie to a dog and you can see a positive response like wagging its tail. You give it a negative stimulus like kicking it and it will exhibit a negative response like barking angrily. We can also condition animals using this theory. In fact, the circus animals are trained in this way. The animals are rewarded when they do what the ring master instructs them to and they would be punished for not obeying the command of the ring master.

It does not work exactly that way with human beings. Human beings are capable of suppressing their immediate impulses and responses. But surely, there would be a response, suppressed for the moment it may be, but lurking just under the surface waiting for an appropriate opportunity to manifest. While all the responses may not be immediate, people respond visibly and positively to a positive stimulus. The alternative is also true, people respond negatively to a negative stimulus albeit not always immediately and in an explicit manner which might be misleading. In most cases, the positive responses manifest almost immediately with words like "Thank you" or something like that. The negative responses usually do not come immediately. In most cases, the negative response comes some time later but the response may not be at a time that is convenient for us or are ready for it. It is very important to take care to ensure that the negative stimulus, when given, is administered in such a way as to generate a positive response. As Mario Puzo says in his novel God Father, "it is not easy to say 'no' and one needs training in administering negative stimuli."

In organizations, we interact with each other all the time. The interactions are much more between a superior and the subordinates. During these interactions, we keep giving stimuli to each other and each of these stimuli would generate an appropriate response. Sometimes, even a negative stimulus can generate a positive immediate response. It can happen in two conditions: one is when the respondent genuinely appreciated the negative stimulus and is willing to correct the situation and the second is when the respondent suppressed the negative response for the time being but keeps it inside to manifest at a time convenient for the respondent.

It is also not possible to keep giving only positive stimuli all the time. Therefore, we must train ourselves in giving a negative stimulus when necessary and when we do give negative stimuli, we must make it appear positive from the standpoint of the receiver of the stimuli.

9.8 Behavioral Correction

Another aspect to motivational theory is the delivery of **behavioral correction** to people who are not putting in performance commensurate with the exigencies of the work, the abilities and the compensation to the person. Behavioral correction is best achieved through counseling and coaching. Coaching is providing expert guidance to the person in how to achieve better performance. This is achieved by providing training (either class room or on-the-job) and expert assistance by a senior person followed by periodic assessment of the progress. Counseling is resorted to when the skill exists but an attitude correction is necessary. The First form of counseling is typically a friendly counseling in which the necessary correction is pointed out in a non-confrontational manner. This tactic can be used when the person is exhibiting the undesirable behavior unintentionally without realizing the consequences of his/her behavior and is amenable to counseling. Confrontational counseling needs to be resorted to when the person is aware that he/she is exhibiting the undesirable behavior even after friendly counseling. We suggest that every manager needs to be to be trained in the art of confrontational counseling. This type of training equips a manager, either through training or self-study, with those skills to better deal with difficult situations. Confrontational counseling involves presenting the expected behavioral correction along with irrefutable evidence of undesirable behavior and the consequential damage. It also involves the outlining the consequences to the person in case the required behavioral correction does not materialize. Confrontational counseling needs management approval to outline the consequences for not bringing about the desired behavioral correction.

9.9 Day-to-day Motivation

Managers have the onus of recognizing the falling levels of motivation and rectify the situation besides elevating the level of motivation in individuals to higher levels and achieve better than planned results for the organization.

Theories are OK and many times, the managers do not have much room to maneuver as the managers do not really have much say in the rewards and recognition or punishment. A performance appraisal may be ten months down the line when the act deserving reward is accomplished today. Promotion may be three years later and then it may be an organizational decision. If someone is performing as expected, then there is no issue but when someone is performing in a less than satisfactory manner how do we motivate him/her to perform better when the time for rewards and punishment is way down the line? This question plagues many managers.

We have a suggested course of action which we used with some success. The following steps aid the managers in motivating their juniors toward better performance.

1. Set fair targets using the organizational baselines and obtain commitments from the concerned individuals.
2. Measure the progress and actual achievement and contrast it with the planned/expected results and derive the variance at pre-determined intervals of time.
3. When the variance is negative, analyze if the individual is lacking some skills and *coach* him/her to improve the skill level so that the performance improves. Coaching need not be handled personally. The individual can be entrusted to a senior person who has better skill in the same area as that of the junior, for improving the junior's skill.

4. We also need to look for any issues, personal or official, that are affecting or inhibiting the achievement of the desired level of performance and remove those barriers to better performance.
5. When we are sure that neither the skill level nor the issues are the cause for the unsatisfactory performance, we need to *counsel* the individual on the need to improve the performance.
6. Then measure again at the next cycle of measurement and repeat the above steps. We iterate until we achieve the desired improvement in the individual.

We utilized this mechanism in our working and achieved good results. We summarize it thus: Set (fair targets) – measure (performance) – analyze (variance) – counsel/coach (if necessary) or simply the SMAC model.

When we attempt counseling, we need to understand that it can turn out to be confrontational counseling. As managers we need to learn about confrontational counseling and be adept at it. First, we need to prove quantitatively that the performance is unsatisfactory not just from the standpoint of the targets assigned but in comparison with the peers of the individual. If we try to subjectively prove that the performance is sub-standard using only our authority, the results would be unsatisfactory. Once we prove that the performance is unsatisfactory objectively, the individual would be amenable to receive the suggestions for improvement. But we must have objective data and be ready to use it. When we use credible objective data, the person would feel that he is shown respect and when we use authority, the person would feel belittled.

When the individual becomes ready for receiving suggestions for improvements, we can coach him/her to raise the performance level. But we need to work on every individual turning in unsatisfactory performance and it may consume significant amount of time. But we need to recognize that motivation is individual and we need to ensure that every one of our juniors is highly motivated. Second, we need to recognize that every junior does not need special motivation as most perform their assignments at a satisfactory level.

9.10 Morale

Morale is also referred to as *esprit de corps* (team spirit). Morale is the resilience of a group of people in the face of opposition or difficult conditions to perform a task or assignment and complete it successfully. It is the capability of a group of people to work diligently toward a common goal setting aside their personal agendas. It is the team motivation, as against individual motivation. It can be observed in team games like soccer, football, hockey, and so on. In team games, individual performances do not win games. True, individual performances contribute to the victory but it is the team performance that really wins games. Morale is the difference between victory and defeat (success or failure) in most instances where a team performance is essential. Morale is the one which can make 1 plus 1 into 3!

We have to note that each member of the team may be highly motivated yet the team morale may be low if they do not work with each other. True, the motivation of each team member ought to be high for the morale of the team to be high but it does not automatically ensure team morale.

The morale of the team is easily noticeable when it is very high or very low. It becomes apparent and manifests itself in the way work is carried out. When the morale is high, the work gets done quickly, with better than acceptable quality and productivity, the issues raised would be very low, absenteeism would be very low, wastage would be minimal, and attrition would be close to nil. When the motivation is very low it would be the reverse in all these aspects. But in most real-life situations, the morale would hover around average, being neither too high nor too low. When the morale of the team is at average level, the outcome would become unpredictable. It would vary with the conditions of the situation. This is a difficult situation for the manager, as it puts the onus of ensuring the output on the manager. It would also be difficult to understand if the morale is going downward or upward. One key feature of very high morale is the pride shown by the people in the work, in defending the team and in covering up for the team members. When the morale is low, there will be a lack of pride in the team members.

Morale can be measured in the team. It is difficult to know the level of the morale when it is at average level. Measurement helps us in understanding the trend of the team morale over a period of time. Measurement removes subjectivity and makes decisions effective and actions fruitful. The methods used in the organizations for measuring the morale are observation and surveys.

Observation of the team performance closely, is used most often in organizations. The results of observation do not yield a quantitative figure though a figure is often assigned by the observer. It is still a subjective figure. The argument of the organizations in defending this method is that morale is itself a subjective influence. It is psychological in nature and cannot be measured and the value obtained by quantitative measurement, itself is subject to factors like faking the answers, inexperience of the surveyor, inappropriate questions/methods and so on. Therefore, many depend on observation method to measure the morale in organizations. This method can be rendered effective if multiple observers are used with each giving a figure on the same measurement scale and a single figure is derived using scientific analysis and statistical methods. At least, this method can be used to determine if a quantitative measurement of morale is necessary.

Morale survey is another method to derive a quantitative measurement of the team morale across an organization or a team. Interviewing, question-naires and analysis of past records are the methods used to quantitatively measure the morale.

Interviewing – this method consumes more time and is costly. In this method, a qualified and experienced individual, interviews the team members either in a face-to-face situation or over phone or video conference. In this interview, the views of the team member are solicited on various aspects of working and rating them on a measurement scale. In this manner, all the team members would be interviewed and a composite team morale metric is derived. The advantage of this method is that faking of answers can be controlled. The interviewer can ask supplemental questions to draw out the person's real intent. The disadvantage is that it is the costliest of the methods of measuring the morale because interviews are time consuming.

Questionnaire – It is a very popular method to measure the morale. In this method a set of questions are formulated and are circulated to the team members either through a paper questionnaire to be filled in or it is put up on the internet and the team members are requested to fill in their answers. Usually the questions have a set of responses out of which, one is to be selected by the employee. This method takes away the requirement of a specialist to conduct the interview and draw inferences. It is easy to collate the answers and derive the figure indicating the morale. The process can be automated both to administer the survey and deriving the figure. It is arguably the lowest cost alternative of the quantitative measurement methods. The big disadvantage is that the answers can be faked. The group can collude and provide a wrong set of answers collectively and deliver misleading the results.

Analysis of the past organizational records – It is not as popular as the questionnaire method. In this method, we look at the records of the organization since the last morale measurement initiative. We look specifically for the attrition of employees, productivity, defects uncovered in the quality control activities, grievances raised by the employees and the customer complaints received by the organization. By utilizing the organizational baselines for these parameters, we can derive a quantitative figure for the team morale. If there is an increase in any of these aspects, there is a deterioration of the morale. This perhaps is the best method as records cannot fake the answers the way human beings can. Men and women may lie but documents don't! We need the time of an expert only for analyzing the results but not to dig up the data. It would perhaps yield the best results for the morale measurement. The cost is medium between the interviewing method and questionnaire method. The flip side of this method is that the organization must be diligent in maintaining records in an easily accessible manner and deriving metrics.

Whatever method is used for measuring the morale, it is important that it is carried out consistently and periodically and a trend graph is plotted. The trend needs to be analyzed and corrective actions need to be taken periodically to maintain the team morale at the appropriate level.

The level of morale would vary over a period of time or with the changing conditions. Some of the factors that affect morale in a team are:

1. The leader or the manager has a strong influence on the morale of the team. The leader can influence the morale by fair/unfair work allocation, rewarding, sensitive/insensitive supervision, critiquing, displaying favoritism, and in many other ways. The team members depend on their leader/manager for facilitation and support in the performance of their activities. The manager can improve the morale by right facilitation and support or damage the morale by not offering the right facilitation support.

2. Induction of a new entrant into the team can affect the morale of a well-adjusted team. The new team member can merge into the team seamlessly and maintain the morale or (s)he could be a square peg in a round hole and adversely affect the morale. Sometimes the new entrant could also fill in a void in the team and improve the morale too.

3. Removal of a key player from the team for any reason including the voluntary departure of the team member can also affect the morale of the team. True, a new player can be inducted into the team but, even temporarily until the team gets adjusted with the new entrant, the morale is likely to be affected.

4. Changing rules or organizational policies could affect the team morale. If the new rules or policies bring in a much-needed relief to the team, the morale could improve. If they are putting extra burden on the team, the morale could diminish.
5. Changing the field of operation of the team could affect the morale. The change could be a new technology or a new geographic location or new methods and tools could affect the morale. Sometimes, they could make the work of the team easier and improve their morale. If they put extra pressure on the team to deliver, it could affect the team morale.
6. Group dynamics are very important in maintaining the morale. Often the team would have informal leaders or opinion makers in addition to the formal leader or manager of the team. When the leader/manager consults the team, these informal leaders emerge and influence the team opinion. These are also referred to as decision influencers. These informal leaders can sway the team morale if they are peeved with any of the factors of team-working.

The sum and substance of the above discussion is that the morale of the team is subject to influences which could be internal or external to the team and can sway the morale. So, it is important that the team morale is measured periodically and the trend is determined so that corrective actions can be taken.

9.11 Rectification of Poor Morale

When we notice that the morale is down from the usual level either through measurement or observation, we need to take actions aimed at correcting the present situation as well as preventing the recurrence in the future. But the first priority is to improve the poor morale.

When we noticed that the morale is down, the first action we need to take is to identify the causes. The morale would never be down mysteriously. There would be some factors building up gradually, affecting the team. We have discussed the factors affecting the morale in the previous section. By careful analysis of the situation, we need to identify the cause or the causes of the poor morale and then take appropriate corrective action.

The first action available to us is counseling the concerned individuals. Counseling addresses the cause of dissatisfaction and removes the misunderstanding or misapprehension, if any, in the individuals. More often than not, the changes induce apprehension and fear in the individuals. This could be due mostly to lack of authentic or due to insufficient information. This could be set right by the manager taking time to explain all the nuances of the change lucidly. Sometimes disciplining one team member may adversely affect the other team members. We need to be transparent in giving rewards which may be positive or negative, especially so when the reward is negative in nature. If there is lack of transparency, then the manager ought to explain to the team the reasons behind such disciplining and set the apprehensions of the team at rest. Sometimes, the counseling may have to be administered to an individual and sometimes, we may need to administer it to the entire team. We need to use appropriate methods and carry out the counseling effectively to restore the team morale to its original state.

There is an adage that a skill, when we fail to practice, would soon be forgotten. So is the case with the goals and objectives of the team. In the humdrum of performing the daily chores, the team becomes jaded and loses its focus due to the monotony of doing similar work in a routine manner. Reorienting basic goals and values at regular intervals could restore the team's perception of their target and keeps them on the right path and restores the morale to its original level. The manager of the team needs to spend time regularly to reorient the team or a team member to focus on the goals and values of the team.

Sometimes especially when a new technology or a tool is introduced, the morale may be down as they find it difficult to cope with the change. Then the team needs to be educated on the reasons as to why this new technology has to be brought in and then the team needs to be trained thoroughly on the usage of the new technology. The team members need education and training whenever new tools, technology or new methods are introduced into the working. Better education and training have shown to improve the morale.

9.12 Preventing Poor Morale

An ounce of prevention is worth a pound of cure – so goes the adage. Rectifying the poor morale is costlier than the prevention of poor morale. It takes time for the morale to slide down before it could be noticed. Then we take some time to confirm that the moral is indeed down and then some more time in rectifying the situation. In the meanwhile, a significant amount of time could be lost while the state of poor morale continues to perpetrate damage. It could cause significant loss to the team and the organization. So, it is better to prevent than rectify.

Manager or the leader has the onus of preventing the deterioration in the morale. The manager needs to keep a tab on the pulse of the team morale to ensure that it stays at the desired level and to notice even the slightest deterioration.

One of the primary actions for maintaining the morale is to measure it at regular intervals. As noted earlier, measurement gives us numeric values which can be subjected to statistical analysis like the time series, extrapolation and trend graphing. These tools help us to identify the deterioration before it hits the alarm button. Once we identify the situation long before it falls below the acceptable level, we can implement the corrective actions discussed in the previous section.

Information is the key to prevent deterioration of the morale. More often than not, grapevine, in the absence of official and authentic information induces unwarranted fear among the team members leading to the deterioration of the morale. Whenever there are rumors, the manager needs to find out the authentic information and pass it on to the team members to allay their apprehensions. Rumors begin when any issue that concerns the employees is not settled quickly and drags on. Large organizations are prone to take time to decide issues. Especially during the time of pay hikes, rumors fly across the organization. The managers also would not have any authentic information on the subject and rumors become unavoidable. Somehow, the organizations have come to maintain a strict secrecy on the issue of pay hikes. We wish that there is more transparency in that area. There are many other areas of organizational working that keep a veil of secrecy in issues concerning employees and this leads to rumors. Secrecy or lack of transparency affects the morale.

The team members need to interact with their managers periodically as it makes them feel that their efforts are being noticed by persons of authority. So, the manager ought to sit with the team periodically and discuss on matters relating to the team. Usually team meetings are organized on a weekly basis. In addition to meetings with the direct manager, the upper level managers also ought to meet with the team on a regular basis but with longer intervals of time. A monthly meeting with the next level of manager would go a long way in maintaining the team morale at desirable levels.

Education and training have shown to allay the misapprehensions of the employees. Therefore, training the team members as required would go a long way in the maintenance of the morale. The term training connotes sitting in a classroom with a teacher lecturing away with zero enthusiasm trying to stuff too much information in too little a time to an uninterested audience. It need not be so. A weekly or monthly knowledge sharing session would also be training. The manager or a member from a different team or the senior member of the team itself could impart training in an informal basis. Infusion of knowledge and information at regular intervals in small doses would be effective in maintaining the team morale.

9.13 Morale at the Organizational Level

While the team morale is high, it is possible that the organizational morale may be low. When the morale of each team is high, the organizational morale can be low. But why? When each team has high morale, the inter-team issues raise their ugly head and they begin feuding with each other instead of collaborating with each other. The team loyalty becomes more important than the organizational loyalty. It usually begins with the team managers feuding with each other and slowly that percolates downward. The interdepartmental conflicts and rivalry are recorded sufficiently to erase any doubt otherwise. It is common understanding that conflict exists between marketing and delivery teams; technical and non-technical teams; HR and finance; quality control and delivery and so on. Most organizations, especially the larger ones, suffer from these rivalries. We cannot avoid conflict in the organizations completely. It is rightly argued that conflict is the key to competition, which raises the bar of performance. But we need to keep it as competition and not as conflict.

But the organization on the whole needs to have a high level of morale not just restricted to the individual teams. But before we embark upon improving the organizational morale, we fist need to measure it. We cannot combine measurement of team morale with that of organizational morale. The individuals may feel that the organization is doing all necessary facilitations needed for better working but yet feel that the "other" department is doing things that are detrimental to the organizational morale. Therefore, we need to measure the morale of the organization by looking at the interdepartmental rivalry and conflicts. Once we do this, we can rectify the barriers to organizational morale and bring about a healthy respect for each other between the departments.

Interdepartmental rivalry and conflicts can be effectively tackled by more effective communication, appreciation sessions about a department's tasks, responsibilities and key result areas and making the interdepartmental transactions transparent.

9.14 Management of Motivation and Morale

As in every aspect of management, the management of morale needs to be carried out at two levels, the organizational level and the individual manager's level.

At the organizational level, the top management needs to provide a framework for the individual managers to effectively manage the morale. The framework includes deriving organizational baselines for the minimum level of morale each team in the organization ought to maintain; provide facilities, or in other words, hygiene factors as stated by Professor Herzberg; provide funding and resources to carry out the activities of morale measurement and analysis; and then encourage the individual managers in achieving the organizational baselines and improving them continuously using a fair reward system. At organizational level, there ought to be an overall plan and schedule for the measurement and variance analysis of the organizational morale as well as for implementing the needed corrective and preventive actions.

Motivation and morale are the manager's primary responsibilities. While the organization provides the hygiene factors for ensuring morale at the minimum level, the individual manager needs to bring in the motivators to improve the morale of the team. The first aspect of managing the morale is the inclusion of this aspect in the team's operational/project plans and record the organizational baseline to be maintained. Then measure it first time and continuously at regular intervals. Then the techniques of measurement, periodicity of measurement, methods and techniques for rectifying and improving morale of the team need to be recorded in the plans.

We also need to provide resources for the activities of measurement, variance analysis and derive the findings. The resources needed may simply be the manager's time or we may engage a specialist to carry out these activities.

Then we need to implement the activities planned, meticulously and diligently. When under pressure to deliver the output specified for the team, we often neglect to carry out these morale building activities. The impact of neglecting the measurement and management of morale may not be felt in the short run but their impact would be very severe in the long run. Often when organizations are small, they would be very focused on maintaining a high morale. But as they grow, the aspects of motivation and morale take a back seat in the frenzy to deliver the revenue-earning output and the aspects of motivation and morale are neglected. This results in unexpected problems cropping up at all places and at inconvenient times. As a sequel, the growth stalls and then it begins to deteriorate. Many organizations that ought to have grown into large corporations do not grow mainly because of neglecting employee motivation and morale.

Motivation and morale are two subjects that are continuously attracting brilliant minds to carry out research at different places all over the world and new realities, methods, techniques and tools are emerging regularly. It pays well for the manager to be abreast of these developments so (s)he can keep the morale of the team at high levels.

10

Performance Management

10.1 Introduction

The sustenance and growth of the organization depends on the performance rendered by its employees. Employees come to work in the organizations primarily to earn a livelihood for themselves and their families. In order to continue in the employment, they would be willing to exert themselves and put in better performance on the job. The only thing the organization has to do is to facilitate and provide support to the employees to put in that better performance. But because there are multiple employees in the organization comparison becomes inevitable. While all the employees would not be performing the same job, there will always be groups of employees performing the same job. For example, in a service organization, work is divided into projects and a team of employees is assigned to each project. One team may be working on projects for certain type of customers, one team working on another customer engagement and one team working on internal projects of similar nature. Each team would have multiple employees doing similar work. While it may not be fair to compare the performance of one team member working in one project with that of a team member from another project, we can compare the performance of the team members working on the similar tasks. When such fair comparison is made, there would be differences in the performances of different employees. The differences in performance could be marginal or large. But at an organizational level, we need to ensure that all performances are closely clustered. In some cases, it is just impossible to bring the variance in performances close to each other. For example, an employee with 10 years of experience in the underlying area would certainly perform much better than an employee with just a year's experience in the same area. Where possible, we need to bring down the variance in the performances of different but comparable resources. In this chapter, let us discuss all the aspects of performance management.

In the days gone by, there was no concept of performance management. An employee performed and if not, he/she was shown the door. But in government and military, they began writing confidential reports as it was not easy to fire people from jobs. The superior wrote a report about the performance of the subordinate and it was kept strictly confidential. This confidential report was used for the purposes of pay hikes, promotions, demotions or discharge.

It was so until late into the 20th century. Then in the late 1970's and early 1980's, there was a change of philosophy about managing people and the name of the department changed from "personnel department" to that of "human resources department" signaling the change in the outlook on how employees were perceived by the management. The change was from looking at the employee as a pair of hands doing what is assigned to human beings to people with thinking power having a stake in the prosperity of the organization as a whole. The word "workers" became an anathema and in its place, organizations began using terms like "resources", "associates" and so on. This change in the outlook led to many changes, the training department became the human resources development department and the confidential report became the performance appraisal. Until the late 1980's, the performance appraisal kept some parts confidential from the employee while revealing some parts. The employee was asked to carry out an appraisal of his/her own performance and the superior needed to compare the self-appraisal of the employee with that of the superior and come to an agreement with the employee. With the dawn of the 21st century, the performance appraisal got metamorphosed with the introduction of the 360-degree appraisal and the periodicity was reduced to a quarter in most professional organizations. Now there is an awakening that the performance appraisal is just a post mortem but not a progressive forward-looking tool. It is not just the matter of appraising the performance alone but it is a matter of managing the performance of our resources. It is in this context, that we are discussing the topic in this chapter.

10.2 Dimensions of Performance Management

As to any other aspect of organizational working, there are multiple dimensions to performance management. These are the dimensions to the performance management:

1. Facilitate better performance management
2. Measuring the performance

3. Performance improvement
4. Performance appraisal.

Let us now discuss these dimensions.

10.2.1 Facilitate Better Performance Management

This is an organizational framework to be put in place by the top management. The top management needs to put in place facilities that would enable managers to better manage the performance of the individuals. First and foremost is the determination of the baselines for various tasks that need to be periodically ascertained for each individual working in the organization. Each individual performance needs to be compared with these baselines and the variance needs to be derived and analyzed. If the variance is positive, that is the measured value is better than the baseline value, we can infer that the employee exceeded the expectation and vice versa. Now the next question is: how many such baselines need to be set for the organization? For each class of activity, we need to have one set of baselines. When we say "set", we mean there ought to be two values out of which one is for productivity or pace of achievement and the second is quality.

Productivity is the amount of work done per unit time. Unit time is usually set as a person day, that is, a person working for one normal work day which usually consists of 8 working hours. The amount of work needs to be measured in some unit for measuring the work. For programming work, we have several measures like the lines of code, function points, object points, software size units, use case points and so on. We need to select one of these units as the standard unit of measure for measuring the size of the output for our organization. Supposing we selected the lines of code as the standard unit of measure for measuring the amount of output, then we express the productivity as, say, 100 lines of code per person day.

Coming to quality, we measure it as the number of defects per unit amount of the deliverable. This metric is also referred to as the defect density. Since we selected lines of code to measure the out put in our above example, let us use that now. The number of defects is just a number. So, we express the defect density as 0.7 defects per 100 lines of code or 7 defects per 1000 lines of code.

So, for a programmer in a Software Development organization, the base line would be 100 lines of code for productivity and 0.7 defects per 100 lines of code for quality. In some other service organizations the base line would be more of a qualitative than a quantitative measure.

In this manner, we need to derive such baseline metrics for each class of activity and provide them to all managers for use in measuring the actual values of their subordinates and derive variance and take necessary corrective action.

One question is pertinent here and that is, how to derive these metrics for the organization. For an organization that has been in existence for some time, we take historical data and with the help of statisticians, we derive a modal value and use it. For a brand-new organization, we borrow these metrics from the industry standard or from another similar organization or an industry association like the IEEE (Institute of Electrical and Electronic Engineers) and use them as our baselines.

One thing is to derive baselines for the organization and the second important thing is to maintain them and make them available to all those needing those metrics. We also need to improve the metrics to reflect the true capability of our resources. As our resources gain experience, their capability would grow and they will be overshooting the baselines. Then we have to derive the metric once again and make them the new baselines. How frequently should we be doing this activity of re-establishing our baselines? It is a good practice and most organizations practice this periodicity which is one year. That is every year, we derive new baselines and roll them out for all to use. This rolling out new baselines usually coincides with the conclusion of the yearly performance appraisals.

In additions to defining and maintaining organizational baselines, the top management also needs to put in place trigger mechanisms to ensure that baselines are derived and maintained and that these baselines are used diligently across the organization as well as analyze the efficacy of the baselines and their usage in the organization.

When we come to supervisory and managerial roles, it becomes very difficult to measure the work and quality. The word managing itself connotes achieving something without full knowledge, facilities and tools. Usually, the managerial effectiveness is measured in terms of the results achieved. It is possible in revenue earning roles. For marketing managers, we may measure the effectiveness in terms of the sales achieved; for software development teams in the value of the deliveries effected; for quality field, in terms of the delivered defect density; and so on. But we have some support departments such as HR department, finance department, planning department, facilities maintenance department and so on. For these people there are no measurable results. These departments support other revenue earning departments. Therefore, we usually set goals such as the finance department fulfills all the statutory requirements on time every time. For these departments, SLAs (Service Level Agreements) are defined and baselined. This kind of service departments are measured by the achievement of SLAs.

In this manner, the top and senior management need to work out performance measures and set up a framework for the operating managers to effectively and efficiently carry out performance management in the organization.

10.2.2 Measuring the Performance

The Hawthorne experiment had brought in a paradigm shift in the thinking of the organizational managements once the results got out and were widely circulated. Slowly but surely, these concepts were not only accepted in the organizations, but also, were implemented in most professional organizations. And, in our humble opinion, the organizations went overboard in their implementation. Thus, the side effect of Professor Elton Mayo's experiments is the feeling it created, that the work study is unnecessary. And, that feeling sidetracked the work study concepts of Frederick Winslow Taylor and Professor Lillian Glbreth.

Lillan Gilbreth

Organizations accorded very low importance and more or less, stopped measuring the performance. Since the Hawthorne experiment proved that the human productivity can be increased irrespective the working conditions, the organizational managements began veering toward the idea that the only path to increased productivity and quality is by taking care of the treatment accorded to the employees and measuring the performance is superfluous.

Working conditions have drastically been improved owing manly due to the two-factor theory of Herzberg, and the treatment of working people improved so drastically, that the present-day treatment has no similarity, whatsoever, with that of the 1970's and the 1980's. While studies need to be conducted to assess the veracity of this new concept, in our humble opinion, it had an adverse impact on the workplace productivity and quality. We have perhaps carried this overdependence on human reciprocation of the excellent working conditions and excellent workplace-treatment by improving productivity and quality. But it did not result in the expected improvement in productivity and quality. While none of the psychologists advocated neglecting the measurement and analysis of the performance, we in the organizations did neglect it. Also true is the fact that the proponents of the work study including method study and work measurement did not advocate neglecting the behavioral aspects. For excellent results, we need to amalgamate and fuse both these concepts and practices to manage and improve the performance of people at the workplace. Hence, this section on performance measurement is included in this book.

Setting up the baselines and an organizational framework for performance would come to naught unless the performance is measured regularly and analyzed. The first and most important aspect of performance measurement is the interval between two measurements. The practice so far has been to do it on a yearly basis. And, we can't call it a measurement in the strict sense of the word. It was for the purpose of annual performance appraisals and it used a 5-point or a 10-point scale or some such other scale. When the superiors rated the performance of their subordinates on this scale, it gave some objectivity to the performance measurement but it is not accurate and there was substantial dissatisfaction in the people, especially the poor performers or super performers. Some people used just three or four levels such as below acceptable level, acceptable level, above acceptable level and excellent level. Some organizations used a few aspects of performance and some organizations used a 25-page booklet. The interval of one year is too long for the superiors to remember or keep extensive records. While it was and even now is being used, one cannot say that it is a good system of performance measurement. The second important aspect of measurement is the factors of performance that need to be measured. Different organizations use different factors for inclusion in their measurement program.

Here is a list of factors that can be included in the performance measurement program, from which one can select the right factors for your organization:

1. **Productivity and quality** – Productivity is the rate of achievement or the pace of working. Each type of work would have different units of work and we need to measure it. We need to recognize that no human being is capable of increasing the productivity beyond a certain limit. One human being's upper limit can be more than the other but each human being has an upper limit. Quality is the absence of defects in the deliverable. We also need to recognize that it is practically impossible to have absolutely no defects when one works continuously, defects creep in due to monotony and fatigue. It is recognized world over that three

defects in a million opportunities to err is to be considered as zero-defect. We measure quality in the delivered defects per unit size of the deliverable. Again, the unit size of the deliverable needs to be defined for each type of work. Now these two factors ought to be included in the performance program for every individual. These two, ought not to be the optional factors. All other factors given below are optional to the performance measurement program.

2. **Goal achievement** – For all positions in the organization, there would be some goals to be achieved. For people with measurable work, goals are the improvement of productivity and quality. For the rest, we set some goals which depend on the specific situation. We measure the achievement of the set goals. To some extent, the measurement of goals achieved is subjective. We need to set, as far as possible, measurable goals. They need to be like reaching a milestone or a delivery to be made, or complete an assignment or something like that which can easily be identified.

3. **Reliability** – This factor refers to the dependability of the person. It is measured in the number of successful assignments as a percentage of the total number of assignments handled. In certain positions, it is more important to complete the assignments with expected quality than productivity. They are usually sensitive assignments like interfacing with statutory authorities, high-value clients/prospects and so on. For people working in logistics, this measure is especially important.

4. **Meeting schedule** – For some assignments delivering on time is extremely important. Filing of reports periodically to statutory authorities is crucial in some cases. In some cases, for example, changing the artifacts in a web site on a specific day is extremely important. In such cases, we measure the performance in the number of times the schedule was met successfully as a percentage of total number of schedules handled.

5. **Economy** – For some positions like purchase function and maintenance function it is more important to save as much money as possible. It is very difficult to measure how much money is saved. A joke is pertinent here: a husband came home panting and proudly told his wife, "Dear I saved the bus fare as I ran behind it". The wife snorted, "You could have saved much more if you ran behind a taxi". But we can use a trend graph of the expenditure over the previous measurement. It can still be misleading as the prices could change and lead us to a wrong impression. So, it is not a good measure to be adopted for performance measurement.

Some organizations do measure economy in resource usage. This is measurable. We can compute wastage as a percentage of total resources employed and then prepare a trend graph.

6. **Integrity** – Integrity is a very important trait in persons with authority and those that have access to strategic information. People in the Research and Development function as well as those in senior management need to have a high order of integrity. It is possible to test integrity in persons. It is also possible to test integrity specific to the position held by the individual. Some organizations do have an entry in the performance appraisal to be filled in by the superior on an impression basis. But there are many instances of people in senior positions sacrificing their integrity for the sake of showing better corporate results or for personal gains. Should we continue to have the method to measure an employee's integrity on subjective basis? We advocate subjecting the selected persons whose integrity is essential to the organization to periodic integrity testing. Then we can use the scores to prepare a trend graph and see the flow. Of course, our organizational finance discipline would have a threshold beyond which the transaction would be subjected to close scrutiny. We must monitor a person's integrity and that would be the function of the person's direct superior. In addition to those, we need to conduct integrity testing of all executives whose function has strategic consequences for the organization.

7. **Completion** – One other factor that is often rated is the completion of the assignments. It is especially in the case of managers handling projects. Some are great at initiating new, high-profile and ambitious projects but leave them midway. It needs someone to bring it to fruition. To identify such people, this factor is rated. It is usually rated as a percentage of completed projects to the total number of projects initiated by the person.

8. **Success factor** – Again this is a factor to be measured for managers. It is measured as a percentage of successful projects to the total number of projects handled. This brings out the capability like no other factor in discovering the overall talent of the person being measured.

9. **Loyalty** – While the use of this factor has comedown significantly in organizations, it is still being used in a few organizations. For sensitive positions like those in the cutting edge of technology either in design or research positions, this factor is used. This is a subjective factor. As yet, there have been no means to measure the loyalty except through psychometric testing which can be faked by intent or by chance.

Like this, there could be other factors that are measured periodically in different organizations based on the unique situation of the organization. Which of the above factors need to be measured is based on the need of the organization.

The third aspect of performance measurement is methodology of measurement and analysis. The most common reason for the performance management is the subjectivity present in the performance measurement. The subjectivity is very high in performance measurement. This is so to give the superior some leverage over the subordinate under the assumption that the subordinates would become lax without the cane-wielding hand of the superior. Well, the cane or the whip was snatched long ago by legislation and this is the remnant of that era. Research by organizational behavior psychologists like Douglas McGregor showed that subordinates are not automatically the reason for any failure. Therefore, progressively, the organizations have steadily shrunk the discretion of the superiors in the matter of performance rating. Organizations have brought in the methodology of measurement and standardized performance measurement as in all other aspects of management. In this methodology, the following aspects are included:

1. Performance factors to be included for each class of employees. In the bygone era, there was just one format/template for rating performance. Now, no more. There are multiple forms for different positions in the organization. We cannot, obviously, use the same methodology for a scientist and a maintenance manager. Each class of persons needs to be measured on the different factors and there could be multiple classes of persons. We need to determine which factors to be used for each class and standardize them. Of course, we can and ought to periodically review the factors in consultation with the concerned and improve upon them either by adding some more factors, modify the existing ones or dropping the obsolete ones.

2. Procedure for performance measurement for each factor needs to be documented and made available to all the employees. This brings clarity to all. The people being measured can do a self-measurement and the people measuring the performance of others have a clear-cut methodology to measure the performance without committing any errors. This brings in transparency and instills confidence in all the employees which augurs well for the organization.

3. Measurement cycle refers to the calendar dates on which the performance measurement is initiated and completed. Performance measurement ought not to disturb any revenue earning work or statutory

compliance work. We need to determine the number of performance cycles in a year and the dates between which this performance measurement must be completed. Most organizations are currently using one measurement cycle per year. We begin it soon after the completion of the financial year of the organization and it would culminate in the pay hikes and promotions. Half yearly performance measurement improves the efficacy of measurement but it is still too long for people to remember the actions taken six months ago. Once in four months is good from both the aspects of remembering the events and at the same time, it gives us three measurements to draw right inferences. Once in a quarter is also similar, but it puts a little more stress on the employees. One cycle provides no data to analyze, two cycles provides two data items but is not amenable to analysis. Three cycles per year or four cycles per year provide enough data to analyze and draw inferences. So, we can select one of these two intervals for carrying our performance measurement in our organization.

10.2.3 Performance Improvement

Performance improvement is essential even if the person performed brilliantly in the last measurement cycle. We need to keep raising the bar continuously as otherwise, the organization would not survive. We also need to identify poor performance to halt and reverse it. We need to identify better performances too in the first stage so that such performers can be groomed to be super performers. After all, the very purpose of performance management is to ensure that all our resources are better performers if not super performers. In our humble opinion, we can achieve super performance from our resources, and what all we need to do is to pay a little attention and devise methods, tools and techniques and implement them without fear or favor. Here are the actions needed for performance improvement.

1. Normalization of data for known reasons and extreme conditions
2. Data Analysis
3. Draw inferences
4. Performance improvement plans
5. Monitoring the performance improvement plans.

   Now let us discuss each of these actions in detail.

Normalization of data for known reasons and extreme conditions – One of the cardinal rules of data analysis is never to accept data as it is received.

We have to normalize the data first by eliminating the data that could be erroneous. Data might have been accurately captured but it could be erroneous because of extreme conditions or known reasons. For example, during the execution of a project, there were several work interruptions due to frequent client-initiated changes or there were heavy storms and the like or some such other issues beyond the control of the team. Using this data, although accurate, would not be conducive to getting the right insights. Similarly, a prestigious project could have been executed using the best resources the organization had. Or a low priority project could have been executed using all trainees. When we consider individuals, a senior person would be guiding and mentoring junior resources while doing the assigned work. Now, if a project needed that expert's time more in guiding and mentoring juniors, then, that person's productivity would take a hit. These cases produce data, that is accurate, but it would not give us the right insights into the results after the analysis. Such data skews up the analysis.

Then the superiors rating the performance of employees would not assign the rating uniformly even after training. Some are too conservative and assign lower rating while some are exuberant and assign a higher rating. We should consider such data also and normalize it by increasing the rating given by a conservative manager and reduce the rating given by an exuberant manager.

Then for each person's rating, we need to take the rating of past two years. This helps us to understand the trend in the present performance of the employee. The trend graph must always be rising. If there is a dip in the trend graph, we need to see if there is an assignable reason. If there is such compelling reason to decrease the performance of an employee, we need to ignore such dip in performance.

We also need to see if the performance has been flat for the last three years. We cannot allow flat lack luster performances continuously. The organization gives pay hikes every year and therefore, the performance must improve every year so that it justifies the pay hike given to the employee.

We also need to see if the performance of the last two reviews and the present review is steadily improving.

After carrying out the data normalization thus, we need to separate such data that has either remained flat over the past three years including the year under review or there is deterioration in the performance.

We need to separate these two categories of employees whose performance has been deteriorating or remaining flat. These are the people for whom we need to draw up plans for improving the performance.

Draw inferences – The data analysis gives us the cases of employees who need improvement. While the data tells us the facts about the performance, it would not reveal "why" or why the performance was so.

So, if a person's performance was flat continuously for three review periods, it is obvious that this person is working at penalty avoidance level of performance. This could be due to:

1. The person is not motivated enough to exert and deliver greater performance
2. The working conditions including the quality of supervision is such that it does not encourage better performance
3. The employee is already performing at the peak and it is not possible to increase the performance any further.
4. The employee has been in fact improving but the superior is doing the employee an injustice by giving lower ratings
5. There is a conflict between the employee and the superior.

We need to uncover the real reason behind the flat performance of an employee. This can be done by discussing the employee performance with the employee as well as with the superior. These interviews need to be done separately so that they can give right answers freely and honestly. In organizations, it does not augur well to go in for confrontations unless it is unavoidable. We need to get information from both and arrive at the real reason for the flat performance. Based on this real reason, we need to take corrective action. The corrective action can be:

1. Put the employee on a performance improvement plan
2. Transfer the employee to another department to work with another superior
3. If any obstacle to performance improvement was discovered, remove that impediment.
4. If it was rated wrongly by the superior, first we need to correct the appraisal and then we need to counsel the superior and scrutinize all the appraisals done by him/her and investigate the appraisal method adopted by that superior.

If a person's rating has been falling steadily for the last three rating cycles, it could be due to:

1. The employee is suffering from a health problem which is preventing that individual from doing his/her best

2. That employee is a misfit in the present position either by qualification, past experience or by attitude

3. The employee could be utterly demotivated due to some reason.

We need to discuss the issue with the concerned individual first about the deteriorating performance to uncover his/her side of reasons and then with the concerned superior about the action implemented so far to correct the performance of the employee. Based on the outcome of these discussions, we have the following alternatives:

1. If there was a health problem, we need to chalk out a plan to improve the health and request the employee to eliminate that issue in a time bound manner

2. If the employee is a misfit in the present position, we need to explore options to transfer the employee to another position which suits the employee better than the present position

3. If there are any motivational issues, we need to ensure to put in measures to improve the motivation level by removing the impediments to the motivation of the employee

4. If none of the above alternatives are available, then we need to give notice to the employee and put that employee on a performance improvement plan. At the end of the performance improvement plan, if the employee does not show the required improvement, then we need to let him go to seek employment elsewhere that suits his/her talents better than the present position.

In this manner, we need to draw inferences about the performance of the employees.

Performance improvement plans – Once the analysis is completed and an agreement between the employee and his/her superior is arrived at, then they need to draw up plans for the improvement of the performance. We need to enumerate all areas that need improvement along with the targets of improvement along with the timelines for achieving those targets. We may use the form depicted in Exhibit 10.1 for capturing the details of the proposed performance improvement plans.

In Exhibit 10.1, the column captioned as "Item Number" is just a running number to capture the number of items needing performance improvement. In the second column, we describe the area needing improvement. It can be brief but it ought to explain the area of the work in which improvement

Exhibit 10.1 Performance improvement plan

\multicolumn{5}{	c	}{**Performance Improvement Plan**}
Employee Name:			Date:	
Supervisor:				
Plan Period: From mm/dd/yyyy To mm/dd/yyyy				

Item Number	Details of the Needed Improvement	Present Performance	Percent Improvement	Final Target

Signature of the employee   Signature of the Supervisor

is needed. In the third column captioned "Present Performance", we need to enter a quantitative (numeric) figure of the present level of performance. In the column captioned "Percent Improvement", we need to enter a percentage by which the present performance needs to be improved. In the column captioned "Final Target" we need to enter the expected level of performance by the end of the plan period. This is arrived at by using the formula:

Final Target = Present performance + (Present performance X Percent improvement)

This plan needs to be reviewed at the end of the plan period. Now, we need to understand two aspects; one is the percent improvement and the second is the plan period. It is tempting to use 5% as the desired performance improvement. It is pointless because the measurement error itself can be 5%! So, depending on the situation at hand, we need to set target as more than 5%. In our opinion, we need to take the gap in the performance levels (desired performance – present performance) and divide it by the number of plan periods in which the gap needs to be bridged. It can be expressed as a formula, thus:

Percent improvement = Percentage of [(desired performance level – Present performance level) ÷ number of plan periods]

We need to arrive at this percentage and round it off to next higher 5 to make it simpler to measure and monitor. It would be a laughing stock if we set 6.82% as the targeted performance improvement!

The second aspect of plan periods also needs to be considered. Yearly performance improvement appraisal is the most common practice in the industry. So, performance improvement plans are drawn up soon after the appraisal process is completed. But if we wait till the next cycle of performance appraisals, it would be too long to take any corrective action in case the target is not achieved. If we have more than one plan period in a year for taking stock of the performance improvement, we would have a chance to correct the performance before the next cycle. In our humble opinion, a quarterly performance improvement monitoring would be effective as the employee as well as the organization have a minimum of three opportunities to correct the erring employee and retain him/her.

Monitoring performance improvement plans – We achieve little or nothing by just making grandiose performance improvement plans if we do not monitor their achievement of otherwise. We need to devise and implement an automated process whereby the supervisors who initiated performance improvement plans monitor those plans and capture the actual results into our database. What we may need to do is to get a software developed to perform these actions,

1. Capture all the performance improvement plans into the database
2. Raise alerts on the supervisors to conduct the measurement and feed the results into the database after getting the concerned employee to agree with the measurement. Better still, if we can get the employee to enter the actual achievement figures into the database.
3. Conduct an analysis to compare the actual results with the planned figures and send an exception report to the concerned executive vested with the authority to take corrective action for the next measurement cycle. The exception can not only be the falling short of the expected improvement but also overshooting the expected improvement. If an employee overshot the expectations in more than one cycle, perhaps, we can take that employee off the performance improvement plan.
4. Repeat these steps at every measurement cycle.

In this manner, we need to monitor the performance improvement plans to bring about real performance improvement in the employee performance. We need to note that the performance improvement is a joint effort between the individual employee and the organization and it is in the interest of both parties to achieve the needed improvement in performance.

10.2.4 Performance Appraisal

Performance appraisals have become a yearly routine and a ritual in almost all organizations. While the original intention of the performance appraisals was to manage performance of the employees at the peak level, it has over time degenerated into an exercise for administering pay hikes and promotions. But progressive organizations have realized that one year is too long a term to effectively manage employee performance. So, they had separated the performance management portion from performance appraisal and are managing performance on a quarterly basis. Still, they are continuing with performance appraisals especially for the sake of promotions and pay hikes. Is this separation desirable? Why can't performance management and performance appraisal be combined? In our humble opinion, they both must be combined. One for performance management and one for pay hikes and promotions look untenable. But having a separate performance appraisal gives some leeway to management personnel to slightly twist the results to do some favors to selected persons who may have to be given higher hikes or promotions due not only to normal performance but due to some exceptional services like winning a prestigious contract and delivering an exceptionally great delivery much ahead of its delivery date and with exceptional quality. Such occasions do occur in organizations quite frequently. After all, the term "management" itself connotes not being able to fit everything into a straitjacket. Management is not something that can be handed over to a computer program based on certain rules. All the same, we need not have two systems for performance management. We can combine both into performance management and accord pay hikes and promotions based on the performance management system coupled with records of special achievement.

In service organizations such as software development organizations, it is common to carry out two types of performance appraisals, namely:

1. Project end appraisals – These appraisals become necessary because an employee may need to work on multiple projects in a year. In such cases, no single project manager would have the full year's information of the employee performance. These appraisals are carried out:

   (a) At the end of the project before releasing the employee
   (b) When the employee is pulled off the project midway due to the exigencies of the organization. The project is not completed at this time
   (c) When the employee leaves the project at his/her choice or leaves the organization or falls ill for a longer tenure

2. Year-end appraisals – these are carried out once in a year before the yearly exercise of pay hikes and promotions. The project end appraisals are utilized in carrying out these appraisals.

Let us now discuss performance appraisals in detail here.

Performance appraisal is the process of assessing, summarizing and capturing the work performance of the employee in an appraisal period. It may be defined as *"a structured formal interaction between a subordinate and his/her supervisor that usually takes the form of a periodic interview (annual or semi-annual) in which the work performance of the subordinate is examined and discussed with a view to identifying the weaknesses and the strengths as well as opportunities for improvement and skill development"*. The objectives of performance appraisals are:

1. Validating the selection of the employee for the assignment as well as other management practices for performing work and achieving results
2. Assisting the employees to understand as well as take onus for their role, results and performance
3. Use the performance appraisal data for making decisions about the pay hikes and promotions.

The purposes of carrying out the performance appraisals are basically threefold. They are:

1. Evaluation of the employee performance – This evaluation needs to be as objective as possible. Objective evaluation gives confidence to the employee about the fairness of the evaluation. It also gives an opportunity to the employee to benchmark his/her performance with those of others in the team/organization and realize where his/her performance is. This realization helps the employee to understand the gap between his/her performance and the ambient organizational performance and to improve the performance.
2. Feedback to the employee – This occasion is utilized by the supervisor to give feedback to the employee not only about the gap in the performance but also on ways and means to improve the performance. It is also utilized to draw up a plan for improving the employee performance.
3. Implement the principle of accountability – Everyone in the organization is accountable for his/her actions and performance. This is the occasion to bring to the notice once again that the employee is accountable for his/her actions. The activities of evaluation and feedback help implement this principle.

The following are the expected benefits from the performance appraisal process:

1. It accords a chance for the supervisor and the subordinate to sit together solely for the purpose of evaluating and receiving the feedback of the

subordinate and to have a detailed discussion of important work issues that might not have had a chance to be discussed.

2. It provides a valuable opportunity to both to focus on work activities and goals to identify and correct existing problems, if any. It encourages better future performance

3. It generally improves motivation level as the employee receives recognition for the accomplishments of the past year. It also brings in more clarity about the role for the next one year. It also provides an opportunity to the subordinate to provide feedback to his/her supervisor and thus the subordinate becomes involved in the decisions relating to his/her work.

10.3 Evolution of Performance Appraisals

Performance appraisals have gone through the following evolution:

1. **Confidential Reports –** These originated in the armed forces in which the superiors prepared a confidential report of the subordinate and filed away in the confidential personal records. No body could access them except by superiors that too at times of selecting candidates for special assignments and so on. The concerned subordinate was never allowed to learn what the boss had written. Every employee was expected to be responsible at the highest level and perform the best. The superior had the divine right to rate the performance as he/she liked without being questioned. Another aspect of this kind of appraisal was that it was free from any format and the superior could write it in anyway he/she feels right. It was not amenable for any kind of analysis. Even the industry and the government used this practice well until the end of the WW II. But the experiments of Elton Mayo and other behavioral psychologists changed the outlook of managements.

2. **Appraisal by superiors only –** The major change after the confidential reports were dumped was to use a formatted performance appraisal that brought in some sort of structure and uniformity into the performance appraisals. This formatted appraisal allowed analysis of the appraisals and draw inferences of how the performance was appraised across the organization. This model gained popularity and was continuously improved to become what it is today. Still the employee being appraised was neither involved in the process nor was he/she shown the ratings.

3. **Appraisal by superior but the ratings shown to the subordinate –** This was the next step of improving the performance appraisal format. The major change was that the employee was merely shown the ratings given to him. It occasionally followed by an explanation by the superior as to the basis for giving those ratings. The employee was usually not given a chance to give his/her side of the rating and if given, the ratings did not usually change. Of course, rare it may be, but in some cases, the superiors allowed employee participation in the rating process and even corrected their ratings. Officially, it was not allowed to correct the ratings. Therefore, the superiors wrote the ratings in pencil and after discussion with the subordinate, they erased the pencil ratings and firmed them up using a pen! But when the superior and the subordinate differed on the rating, the decision of the superior prevailed with the subordinate having nowhere to appeal his/her case.

4. **Appraisal by superior but the subordinate was given an opportunity to enter explanation or his/her version –** The next improvement was to acknowledge and formalize that the subordinate has a different perception of the ratings. Against every rating awarded by the superior, some space was provided to the subordinate to write in his/her version of the rating. Initially, the subordinate was encouraged to simply agree with the superior's rating and all that could be written was "I promise to improve". But slowly, the subordinate was allowed full freedom to express his/her version freely.

5. **Self-appraisal along with the appraisal of the superior –** The final improvement, till now, is to allow the subordinate to fill in his/her appraisal along with that of the superior. In every aspect of rating, a space was provided to the employee being appraised to award his/her own ratings and explanation of those ratings. The superior also has the same opportunity. Initially, both were encouraged by the managements to come to an agreement such that both ratings are same. The better bargainer prevailed. In some cases, the subordinate could convince the superior and, in some cases, the superior convinced the subordinate. Some organizations recognized that an agreement between the superior and the subordinate is not essential. In those organizations, they provided a separate space at each rating in which the superior can record the reasons for the difference in the ratings and this explanation is not shown to the subordinate usually. Currently, this is the method that is used in most organizations.

6. **360-degree appraisal** – With the onset of the capability maturity models, beginning with the CMM (Capability Maturity Model) of the Software Engineering Institute of Carnegie Mellon University in 1998, which literally flooded the market. Now there are more than 30 varieties of capability maturity models in the market. One of them is PCMM (People Capability Maturity Model) which is applied to assessing the capability maturity of the people working in the organization. One of the concepts of PCMM is 360-degree appraisals. However, the concept of obtaining feedback on employee performance from peers and subordinates in addition to superiors predates PCMM. PCMM gave a concrete shape and formality to this concept. An individual working in the organization is responsible to his/her superior – true but the individual impacts not just the superiors but also his/her subordinates and colleagues. It is possible that the employee ensures that his/her superior is satisfied with the performance but the employee may be adversely impacting the performance of his/her colleagues. This needs to be detected as the organization needs all employees to be working harmoniously to achieve over all productivity and quality of the deliverables. So, it makes sense to obtain the feedback from the employee's peers to get a better view of the employee performance. Similarly, the superior may be adversely impacting the morale of his/her subordinates, which may impact their performance and morale over a longer period. Therefore, it also makes sense to obtain the feedback on the performance of the employee from the concerned subordinates too. Thus, obtaining the feedback on the employee from all the four directions, namely, superiors (from above), the subordinates (from below), the peers in the downstream activities (left) and the peers from the upstream activities (right) completes the circle and hence called the 360-degree appraisals. While many progressive organizations are practicing this concept, the actual implementation differs from organization to organization. Implementing a full-scale 360-degree performance appraisal system consumes significant amount of time both for getting them done and later analyzing them too. So, many organizations do not implement the concept in toto.

10.4 Content of Performance Appraisals

The performance appraisal normally contains facilities for entering objective data as well as narrative description. The objective rating uses a rating scale which is generally a 5-point scale or a 10-point scale. There are but rarely, a

100-point scale. In 5- and 10-point scales, a verbal description is added for each point but in the case of 100-point scale, no such verbal description is used but only a number is to be entered. In 5-point scale, descriptions like Excellent (5), Very Good (4), Good (3), Fair (2), and Poor (1) or some other equivalent captions are used. What we need to understand is that 5 is the best performance, 4 is above average, 3 is average performance, 2 is below average and 1 is unacceptable performance.

What aspects should be included in the performance appraisal is a question to be answered by the individual organization depending on their unique situation. It would be advantageous if more objective aspects are used as objective data is amenable to analysis than subjective aspects which cannot be subjected to analysis. While we cannot prescribe aspects to be included in the performance appraisals which are universally applicable, the following a list of recommend aspects for inclusion in the performance appraisal:

1. **KRA achievement** – This is applicable especially in the managerial cadre. Each manager would have some KRAs (Key Result Areas) that need to be achieved. This is the minimum performance expected of a manager. This would be a subjective measure. So, we need to provide space for narrative description by the superior along with a space for the subordinate to offer his/her version of the situation.
2. **Productivity** – This is an objective measure. It is the rate of delivery or the pace of working of the individual.
3. **Quality** – Quality is the absence of defects in the delivery effected by the individual. It can be measured by the number of delivered defects per unit size of the deliverable. In software development it is expressed as the (number of defects per 100 lines of code).
4. **Schedule** – schedule is delivering the deliverable on or before the agreed date. The individual is expected to carry out assignments in the appraisal period. So, we just count the number of times the schedule is met and compute the percentage of meeting the schedule commitment. Now what about the number of times when the individual beat the schedule, that is, the employee delivered even before the schedule date? We need to give special credit to that exceptional performance. So, we need to include the percentage of times the schedule is met, the percentage of times the schedule is missed and the percentage of times the schedule is beaten to give a fair rating of the performance of the employee.

5. **Dependability** – Dependability is both a subjective as well as objective rating. Dependability is the capability of the individual to accept and complete an assignment even in adverse conditions. When a person refuses to accept an assignment even when it falls within the person's skills, that person becomes undependable. When a person leaves a task midway without completing it even when he/she has the required skills, that person becomes undependable. We can express this as a percentage, the number times; the individual either refuses an assignment or leaves it incomplete to the total number of assignments entrusted to the individual including the ones refused by that person. But it would not be fair if we do not allow the individual a chance to explain the situation. So, we need to include a space for the explanation of the individual being appraised.

6. **Teamplay** – This is rather a subjective measure. So, we may need to provide a space for the superior to list out the instances of affecting team work and a space for the subordinate to offer his/her explanation.

7. **Handling stressful situations like urgencies** – Stressful situations are part of work life in any organization. This can be subjective as well as objective measure. We can express the number of stressful situations in which the employee did not meet the expectations to the total number of stressful situations faced by the employee. We need to include space for including the explanation of the superior as well as the subordinate to offer their respective versions of the failure.

8. **Integrity** – Integrity is extremely important for sales people and senior managers. Lack of integrity in these people can cause serious damage to the organization. Most progressive organizations conduct some type of integrity testing while recruiting these people into the organization and also periodically to ensure that these individuals continue to exhibit integrity in their actions. While so, we need to note that it is extremely difficult to prove lack of integrity in an individual. Some organizations allow a narrative description for this measure with provision for the individual to offer his/her version. Some organizations leave this aspect out of performance appraisals but conduct integrity testing for their employees once a year or so.

There could be other measures or aspects of performance that may be included in organizations depending on their unique requirements and situation.

10.5 Errors in Appraisals

Errors do creep into the performance appraisal in spite of diligent implementation of the process in the organization. Here are some of the reasons for those errors:

1. Poor record keeping – to carry out effective performance appraisals, the superiors and the subordinates need to keep impeccable records of the performance of individuals. When the superior's ratings are favorable, it causes no friction but if they are unfavorable, it causes friction. So, it can be resolved amicably if there is an irrefutable record of the performance leading to that rating. Errors creep in when such records are not kept.
2. Recent event effect – As the performance appraisal covers a year, the events that took place in the distant past would be pushed back in our memory and the recent events come to the mind first. If that recent event is of exceptional significance, then it dominates the rating process. If that event resulted in negative impact, it is possible that all performances of the individual can be rated negatively. If that recent event resulted in significant positive impact, then all mistakes of the individual would be forgotten and every aspect is likely to be rated positively. We need to avoid such impact to be objective in our appraisals.
3. Prejudices – All of us suffer from our own prejudices because of a variety of reasons. We need to recognize this fact and take special care and put in efforts to rate the performance based purely on the performance of the individual and the records we meticulously kept about such performance.
4. Disproportionate ratings – Some of us suffer from this effect. Some are prone to award a bit better rating and some of us award a bit lower rating. It is an inherent characteristic. We need to train ourselves to come out of this kind of rating and award accurate ratings to the individual being appraised.
5. Not allocating adequate time for this activity. We are prone to procrastination resulting in not having adequate time for carrying out an effective performance appraisal. This leads to errors in the ratings awarded in the performance appraisals.
6. Lack of senior management support watering down the seriousness of the process of performance appraisals.

7. The process for ensuring that the performance appraisals may not have been defined at all, leaving it to the people which will result in poor quality performance appraisals.
8. If the process was defined, it may have been poorly defined or not improved with the changing conditions resulting in lowering the quality of the performance appraisals.

There could be other situation-specific errors that might affect the accuracy of the ratings awarded in the performance appraisals. We need to take every care to ensure that the performance appraisals are accorded due importance and are taken seriously in the organization and that they are free of any inaccuracies.

10.6 Prerequisites for Effective Performance Appraisals

We need to put in place an efficient framework using which the managers can ensure that the activity is carried out diligently. Here are some of the factors that we need to put in place before we initiate the practice of performance appraisals.

1. Define a proper job description for each role in the organization covering the role, responsibility, accountability and the KRAs.
2. We need to set performance targets in as objective terms as possible at the beginning of the appraisal period so that each individual is aware of what the organization expects from him/her.
3. Define the executives who would be carrying out the performance appraisal and the manager who would review the completed performance appraisal.
4. We need to define the procedure for carrying out the performance appraisal and conduct a training program for all appraisers so that they would carry out the performance appraisals in a uniform manner
5. We need to standardize a scale for measuring the performance of the employees. It could be a 5-point or 10-point or a 100-point scale but we need to select one for the organization and standardize it.
6. Define an escalation mechanism when the superior and the subordinate refuse to agree with each other. Usually, it would be the reviewing manager but we can also define a separate mechanism to resolve appraisal disputes.
7. We need to standardize the interval between two appraisals. In most organizations, it is one year, but in progressive organizations that focus

more on performance management, it is quarterly. Whatever it may be, we need to standardize the interval to remove any ambiguity.

8. It is not only the appraisers that need to be trained. We need to train the appraisees also. They need to be trained on record keeping to get the rating they deserve, how to negotiate better rating and finally how to escalate the issue when they do not agree with the appraiser.

9. Then design the performance appraisal forms appropriate for the organization and obtain approval from the top management for implementation in the organization

10. As we noted in the earlier section that appraisers can be conservative and award lower ratings or too liberal and award unduly higher ratings. We need to normalize these ratings. We need to define a process for normalizing the ratings. This is usually carried out by correcting the rating, given by the managers with a known tendency for being either too liberal or too conservative. Second, we also normalize these ratings based on the success of the projects handled by the appraisees. If a project is hugely successful, we cannot award lower ratings to any of the people that worked on that project – right?

These preparations make us ready to conduct effective performance appraisals in our organization.

10.7 Quality Assurance of Performance Appraisals

Without credible quality assurance any human endeavor is likely to degenerate into a wasteful activity. We need to subject every activity in the organization to its applicable quality assurance activity. We also refer to this as a system of checks and balances. The first quality assurance activity is carried out at the source, that is, immediately after the performance appraisal is completed. In most organizations, as soon as the appraiser submits the performance appraisals, they will be reviewed by an executive superior in designation to the appraiser or a by a specialist from the HR department. The purpose is to verify each item in each of the completed appraisals to ensure that it is indeed filled in, filled in correctly and that it conforms to the organizational guidelines in every aspect. The verification also includes the activity to ensure uniformity among the appraisals and that the ratings are neither too liberal nor too conservative and they are having parity with the ratings of other appraisers. If the reviewer finds any deficiencies, they will be sent back to the appraiser for rectification.

The second quality assurance activity that is carried out is periodic audits. Audits are basically of two types. One is conformance audits and the other is investigative audits. Audits are primarily document verification systems that verify the records maintained while performing the work to ensure that all activities are performed conforming to the existing procedures and that all approvals had been duly accorded. Conformance audits are carried out periodically or at the conclusion of a significant step to ensure that every thing was performed as specified. Investigative audits are performed to uncover the reasons behind an exceptional event. The exceptional event can be a major failure or a grand success with a view to uncovering the root causes so that future failures can be prevented or the present success can be replicated in future.

In most progressive organizations, immediately after the performance appraisal activity is completed, a conformance audit is conducted. Normally this audit is conducted on the HR department to ensure that the process has been initiated on time; that all concerned were trained appropriately; that adequate time was allowed to the managers to complete the appraisals; that all appraisals were reviewed; and that necessary normalization of ratings was carried out. If the audit discovers any non-conformances, they will be given to the HR department which takes immediate corrective actions to correct the non-conformances and to put in place necessary preventive actions in the process to prevent recurrence of the errors in the future appraisal process.

In this manner, it will be ensured that the process of conducting performance appraisals was carried out diligently and that quality was built into the process.

10.8 Process for Administering Performance Appraisals

For any activity to be carried out efficiently and effectively in an organization, the management and the people need to work together shoulder-to-shoulder cooperating with each other. Both have their roles and responsibilities, which they need to carry out efficiently and effectively. A well-defined and continuously improved process is the framework put in place by the management to be used by the people in the organization so that the desired results would flow from the activity. A process consists of the following:

1. A process which is the top document that gives the overall intent of the document, objectives, agencies involved and calls all other documents
2. A set of procedures to perform intended activities like, initiation of the appraisal process, execution of the appraisal process, closing of

the appraisal process, rating scale, rating procedure, review procedure, rating normalization procedure, analysis procedure, audit procedure and so on

3. Formats and templates which include a template for recording the performance appraisal, collation of completed appraisals, formats for review comments, analysis formats, non-conformance report for recording audit findings, and other related formats
4. Checklists – Checklists help to ensure that an activity is completed comprehensively. A checklist contains a number of items enumerated with questions about the activity that is being carried out or verified. Each item can be checked off if completed. When the activity is completed, all the unchecked items in the checklist show the pending or unperformed actions in the activity. Our process would have checklists for completing the appraisal, verification of the appraisal, performing the analysis, verifying the results of the analysis and so on.

Any system or process or procedure undergoes change as time elapses due to changes in the environment, business requirements, governmental regulation, technology and the exigencies of work activities. So, the process needs to be upgraded periodically to handle all these changes. But we need to do it conforming to a process improvement process. It would contain procedures for the agencies that can raise a process improvement request, the procedure to handle all the process improvement requests, the procedure of analyzing and selecting the requests for implementation, the procedure for changing the applicable artifact, getting it reviewed and approved and finally changing the artifact in the process library with the upgraded artifact. It would also contain the format for raising the process improvement request, its authorization, analysis as well as applicable checklists to ensure that all the applicable activities were completed comprehensively. Using this process, we improve the process periodically and ensure that the performance appraisals are performed in the organization diligently, comprehensively, uniformly with built-in quality.

11

Skill Retention and Development

11.1 Introduction

Before we talk of skill retention and development, we need to remember Laurence Peter and his Peter Principle. Peter Principle was developed by Dr. Laurence Peter. He published it in a book by the same name which he co-authored with Raymond Hull. He stated thus, "In a hierarchy, every employee tends to rise to his level of incompetence." He continued, "In time, every post tends to be occupied by an employee who is incompetent to carry out its duties. Work is accomplished by those employees who have not yet reached their level of incompetence." This work has, in our humble opinion massively changed the outlook of the managements and they brought in what was initially referred to as "training", which progressed to "training and development" and now to "human resources development". But for this work and publication of the Peter Principle, it would not be an exaggeration to say, that we would not have had the skill retention and development we have in progressive organizations today. In that book he talks of the situation in a repair shop. They promote an excellent mechanic to the post of foreman of the shop. Now, every mechanic called upon the foreman whenever he/she had some trouble and handed the job to him. While he worked on that problem, the rest of the mechanics took it easy. In no time at all, the shop situation deteriorated significantly. The book gives much more information and support to establish this principle.

We have a duty to ensure that nobody reaches his/her level of incompetence in our organization. That is why we conduct various programs of skill development. We also conduct some sort of evaluation before promoting people to the next level to ensure that the individual is indeed capable of performing the duties assigned for that level.

But, why should we spend efforts in the skill retention of our employees? They are doing the job every day which ought to retain their skills – right? Very true in manufacturing industry where the products change but not the production process or tools. But, in the service industry, not only the products change but also the tools, the methods and the way we build the deliverable change, especially in industries like the software development and other similar service industries. While the change may not be every day, it would be there once every three years or so. New tools are coming on to the market every day, new methodologies are hitting the market almost by the day and new clients, demanding a different variant of the service, are coming our way. If you consider software development industry, one client may ask you to conform to the ISO 9000 series of standards and another may ask conformance with SEI-CMMI© and yet another may ask you to implement Six-sigma methodology. We do not know what tomorrow would bring. The hallmark of the 21st century has been of faster obsolescence. Nothing is stable any more. A lot more research is being conducted on every aspect of human life including work life and new discoveries are being made every day. In service industry we need to look at all these and become ready to handle the changes. Of course, we do recruit people of higher skills with faster learning capability but we need to make sure that their skills are sharpened, honed and upgraded to do their present job efficiently besides making them ready to handle tomorrow's jobs.

We will discuss this topic in this chapter.

11.2 Skill Retention

Why is there a need for retaining the existing skill? It is enough if it is practiced – right? Right. What we meant by "skill retention" here is to make sure that the employee has the right skill to perform his/her present job efficiently and effectively even when the work environment, methodology and the tools change and they undergo change periodically. So, while the nature of the job to be performed and the deliverable remains the same, the skill needs to be adjusted to perform the job without affecting the efficiency or the effectiveness using new tools, new environment and methodologies. Now, let us take a software tester, who was performing manual testing. Now the organization, decided to introduce tool-based testing. Now, the tester is carrying the same old software testing and delivering the same test report but with the help of a testing tool. When the organization introduces testing tool and trains the tester to use the tool, will it be skill-upgrade or skill-retention? In our humble opinion, it is skill retention because the individual is doing the same job of testing but using a different tool. In manual software testing, it was not that no tool was used because it was carried out with the facilities provided by the operating system. The important aspect to note here is that the *individual is not trained to do a different job*! The individual is doing the same job but in a different way, perhaps using a different tool or a different method.

Therefore, skill-retention is to retain the skill to perform the same job in changed environment at the same level of efficiency and effectiveness.

To retain the skill, we do take number of actions in the organizations. They include apprenticeship under a senior person, on-the-job training and a structured training program. We will discuss about them in the coming sections.

11.3 Skill Development

While skill retention helps us in continuing to do our present job without any loss of efficiency or effectiveness, skill development gives us the ability to do our present job with better efficiency and effectiveness or perform the duties of the next level job efficiently and effectively. Most progressive organizations would have a career plan for their employees in which the employee progresses from the initial job and moves up on the corporate ladder provided that individual performs his/her job as expected producing the expected results at regular intervals. But before an employee can be promoted, the organization ensures that the employee is not affected by the Peter Principle, discussed above. Therefore, we usually conduct some sort of skill development program for the individual just before or immediately after the individual is promoted. While the skill retention is primarily focused on imparting the required technical skill, skill development is focused on the technical skills, soft skills and the leadership skills.

Soft skills are focused on making the individual a better person to make him/her a better team player. They include such topics as interpersonal relations, effective communication, group dynamics and so on. Leadership skills equip the participants with the necessary skills to get the job done by others, as well as developing, coaching and mentoring the subordinates. Leadership skills programs would also include some amount of management theory, techniques and tools depending on the level. Leadership skills programs would normally be at three levels. The first level is focused on those people who are at working level to make them suitable for the first level supervision. These programs are usually referred to as supervisory development programs. The second level would be to equip the first level supervisors with necessary skills to become middle managers. These programs are usually referred to as executive development programs. The third level program is focused on developing middle managers into senior managers. These programs are usually referred to as management development programs.

11.4 Vehicles for Skill Development

Vehicle is the mode of imparting the training to the participants, some of which are enumerated here.

1. **Self-study** – The individuals are encouraged to use self-study of the manuals like the user manuals, troubleshooting manuals, administration manuals and so on and learn the way the tool works and gain mastery over it. It is sometimes the only way, especially when we import the tool from a foreign country or from a long distance in which case, it will be very costly to either depute our people or to get the vendor expert to come to our place and train our people. This does work very effectively in most cases where the people involved are well educated and are capable of studying and learning from the manuals. Now it includes, in addition to reading manuals, watching videos where available or listening to podcasts if available. Most organizations use this training technique. However, we need to give time to the persons involved for studying and discussing among themselves to clarify difficult portions and to do some amount of experimentation or trial and error to become working level experts in the subject. However, this is being used diminishingly because of the technology advancements. Still, self-study is a significant skill-retention method especially for individuals when structured training is not available. This is used especially when there is not adequate number of candidates to arrange for a structured training program. In a structured self-study program, a syllabus is defined, the suggested/prescribed study material is pointed out and quizzes or questionnaires are prepared for the evaluation and in some cases a structured evaluation is also implemented.

2. **Web-based training** – Now with internet, it is not necessary for people to be in one class room to impart or receive training. The trainer can remain in his/her normal working place and impart training to a set of trainees sitting in a class room, thousands of miles away, looking at a large screen and listening from the speakers or headsets. Even better, the participants can be geographically distributed looking at their laptop screens and listening from headsets. The participants can ask questions and get answers. We can even conduct tests for widely distributed

participants from a central location. It has proven to be very effective and is slowly replacing the in-person class room training to a large extent. We can confidently say that this method is the most popular in the present day.

3. **Video-based training** – This has become common place in the recent times. Many videos are now available in web sites like the YouTube showing how to perform certain actions. It has also become common place to video tape the specialist training programs and use them later to train another batch of trainees. These videos are also being used to train individuals. The advantage of these videos is that the candidate can learn the subject at his/her own pace and convenient time. The professional training institutions produce and make these videos available to organizations at affordable prices. The cost of these videos would be much less than that of hiring a faculty and conducting the training program.

4. **Knowledge sharing** – When an organization introduces a new programming language or a testing tool or any other major tool, they usually organize an introductory training program to impart necessary skills to the users. Any training of this nature is akin to the training to teach driving automobiles. The trainer teaches enough to the budding drivers to use the vehicle and to take it through the traffic without causing accidents. But to become an expert driver or become a racing driver, one has to gain it by practice. So is the training for tools and programming languages. It will be adequate train the people to handle the tool and to get the work done. But to use the tool to its full potential efficiently and effectively, we need to use it extensively. But everyone will not get the opportunity to use the tool extensively and on every aspect. So, the concerned people conduct a knowledge sharing session, in which the tool users will share their experiences, the troubles they faced and how the surmounted those troubles. In this way, everyone gets the benefit of every other's experience and knowledge. The collective experience and knowledge, becomes the knowledge and experience of each team member. In most progressive organizations, it is mandatory for each team to conduct a knowledge-sharing session periodically in which the team members share their experiences. This leads to more effective and efficient performance on the present job of every employee.

5. **In-company training programs** – In company training programs are conducted when there are sufficient number of candidates needing training on the same set of related skills that can justify the cost of

conducting a structured training program. Conducting an in-company training program is a costly matter. We need to arrange for a classroom, training facility, faculty time when we use internal faculty or cost of hiring an external faculty member in addition to administrative and coordinating costs. One advantage of conducting an internal program is the feasibility of adding more participants if necessary. External programs would charge "per participant" fee to admit participants. In internal programs, the costs are fixed irrespective of the number of participants and those costs can be better utilized by adding more participants which will reduce the "per participant" cost. Another advantage, of internal training programs, is that we can have control over the course content, the duration and the way it is imparted. We can extend the program by a day or two depending on the situation, if required. We can also revise the course contents midway through the training program. We can also sponsor participants for specific topics of the training program rather than for all topics.

6. **Sponsorship to public training programs** – Professional associations such as the IEEE and research bodies like universities and consultancy organizations conduct professional training programs for dissemination of the latest research or developments in the field to the practitioners. They would also conduct training programs in fields in which there exists a knowledge gap or the field needs more people. They design the course including the course contents, duration and method of conducting the course at a fixed fee per participant. Sometimes, they give a bulk discount to organizations that sponsor multiple participants. Most professional organizations take advantage of these programs as the latest developments and the best practices culled from many organizations would be included in these programs. Such knowledge is not available without paying a hefty fee for technology transfer. These public training programs make such knowledge available at a very low cost to the participants. These organizations announce these programs publicly as well as privately through mailers including emails. We can sponsor as many candidates as desired by our management. One advantage of these training programs is that they are very cost-effective. The other advantage is that we get access to latest developments in the field which are not otherwise available to us. Sometimes these programs may not be available in our city and we may need to send our people to other cities incurring extra costs such as travel, boarding, lodging and local transportation costs. One precaution we need to take is that we need to

ensure by formal methods that the knowledge acquired by our employees is disseminated by the participants to others in our organization using a suitable method like knowledge sharing. Otherwise, the acquired knowledge may walk out of our organization when the specific employee leaves us for any reason.

7. **Sponsorship to seminars** – The faculty member needs to use the technology for some amount of time and gained a reasonable expertise, before conducting a training program on the subject. The fact that someone can conduct training on a specific topic, is an indication that the technology is already being used by someone and that others had implemented that technology and stabilized it. The very purpose of training, either with internal faculty or external faculty, is to catch up with the leaders in the industry. Training programs do not prepare our organization for the future. Training programs are OK for organizations in the second or third rung from the bottom of the industry leadership ladder but not for the industry leaders. Therefore, we need to use another vehicle that would prepare us for the future to retain our leadership in the field. Seminars are the answer for this question. Seminars are always futuristic. Seminars are usually conducted by professional associations or universities. While the seminars conducted by professional associations are focused on knowledge sharing of the latest developments in the field of applied research, the seminars conducted by the universities are focused on sharing the latest research findings of basic research. Both are useful for organizations. The proceedings of the seminars conducted by the professional associations can become immediately useful and be implemented in the organization, if they are in our line of business or can be dovetailed into our practices seamlessly. The proceedings of seminars conducted by the universities, on the other hand, are completely futuristic and will give us hints on how the future developments are going to be. These will help us to strategize our long-term plans. Therefore, we need to scan the seminars being conducted by various professional associations and the research universities and shortlist those that are in the same field as our organization and then refer them to the concerned executives in our organization to see if those seminars are really useful for our scenario. This is a collaborative effort between the HR department and the technical department. Most executives in the organization would be members of a professional association of their specialization and receive the publications of that association. Such publications would contain the details of the seminars and training programs

being conducted by them. They can then select the relevant seminars and bring them to the notice of the HR department which will take it over and complete the required formalities to sponsor our employees to that seminar. Universities would also inform organizations using emailers and brochures from which we can select relevant seminars useful for us. Seminars are usually charged on a per-participant basis but would allow discounts for bulk sponsorship. One disadvantage of sponsoring to seminars is that the actual programs may differ from what is advertised. Sometimes, the papers submitted in the seminar may be of poor quality. At the end of the day, the employees sponsored by us may just leave the organization. We can never assess the effectiveness of the seminars fully.

8. **On-the-job training** – This is the most frequent skill retention and development tool used in the industry. This is especially applicable in the case of experienced resources. When a working level person is promoted to a senior position in the working level itself, the person would be trained on the job for the additional responsibilities, incumbent on the new position. The additional skills needed for the senior position are slightly higher than the present position and these would be imparted to the individual while he/she is working on the new job. On-the-job training would be very effective in smoothening the rough edges of performance and for fine tuning the skills. This is also utilized in inducting a newly recruited person into the position. The newly recruited person would have all the necessary skills but is new to the environment. The familiarization of the environment to the newly joined employee would be carried out while the individual is performing the job. In this method of skill development/retention, the supervisor of the employee as well as his/her colleagues would impart the necessary training either by demonstration or explanation of the way to perform the job. There would be no set curriculum or evaluation of the impact of the training program. There would also be no set duration for the completion of the program. The employee learns the needed skills at his/her own pace or as dictated by the environment.

9. **Subscription to journals and magazines** – Most professionals working in organizations would be members of professional associations which publish a journal that would be mailed to the members free of cost. These journals could be print versions or e-versions. The share of print versions is dwindling steadily and in the near future would be extinct in favor of e-versions. In addition to the journals published by the

professional associations, business organizations, some universities also publish professional and scientific journals which publish peer-reviewed high-quality articles, viewpoints, interviews, latest developments, new product reviews and so on. All such writeups would be penned by experts in the industry or academicians and renowned consultants. They would provide much value to the reader in terms of providing glimpses into both the present and future of the industry. Such commercial publications would be costly for individuals to subscribe to. Therefore, organizations would subscribe to such relevant journals and provide them to their employees through the library. The employees can borrow those journals from the library, study them and update their knowledge. Of course, the library would be stocked with latest books on relevant topics along with these journals. The library is the heart of the knowledge management initiative and these journals are a vital part of the knowledge management in the organization.

10. **Snippets circulation** – One more aspect as part of knowledge management is the circulation of snippets culled from various journals. All employees are always kept busy by their usual routine and in meeting their targets. They would have limited time to read journal after journal to cull what is relevant for them and study what was found relevant. Therefore, the organization would designate a few employees to just prepare abstracts of articles or papers relevant to the organization and then circulate them to concerned employees. Generally, the journal is abridged to one or two pages containing the headlines and a line or two of explanation what to expect in the full article. The executives would look at that and borrow the journal or article if it is of interest to them. Most professional organizations use this practice to ensure that their executives have access to the latest developments so that they utilize that information in their working. In the present days of internet, these snippets come in e-form in email or bulletin boards with links to the full articles.

11. **Podcasts** – This is another recent trend that is fast catching up. Now that internet has reached every nook and corner of professional organizations as well as the developed world bringing great speeds to stream audio and video, this technique is fast catching up. Producing a professional video is costly because it involves professionals who need to speak into the camera. Professionals are not trained actors to look impressive on screen. Videotaping needs studios to prepare the sets for shooting the film or work places to be tidied up to look clean and presentable in

the video. People without make up do not look appealing in the videos. While it is technically possible to video tape people at geographically disparate locations, the quality would not be that good. Audios on the other hand are relatively much cheaper and easier to produce. We can produce audios of good quality with different people at different geographical locations. Free tools and low-cost tools are available to record audios of people in different locations. It is also comparatively cheaper to edit audios than videos. Therefore, audios are being produced explaining a concept or a development and interviews and are being circulated. It is also easier to listen to audios during free times or while traveling on a bus or a train. Now every smart phone is equipped to play audios with earphones. Many professionals and professional organizations are producing these audios and are circulating them to those that signed up with them. These news letters with audios are being called podcasts using the name of the Apple product iPod which was the first to popularize listening to Internet audios. Now there are umpteen websites that provide these podcasts to listen directly or download and listen freely. www.spamcast.net is very popular web site providing about 500 podcasts that can be listened free of cost!

12. **Membership to professional associations** – Almost every field of human endeavor, there are associations of professionals whose avowed purpose is to disseminate knowledge of that field as well as to develop its members in their field of activity to the betterment of the industry and thereby the society. They conduct lecture programs, training programs, and seminars besides publishing a journal dedicated to the profession. They bring out standards and guidelines to streamline various activities as well as to build in the best practices into the activities and ensure a minimum level of quality into the deliverables. The associations allow hefty discounts to their members in all the programs they conduct. They also define a code of conduct for their members in their practice of the profession. In some professions like the practice of medicine and law, they also test and certify/license a person to practice the profession and the association would also decertify/delicense a member from practicing the profession too. To continue in such professions, it is mandatory to update their knowledge and produce proof thereof to the association to continue their certification/license. While it is compulsory to be a member of the relevant professional association in certain fields, it is not mandatory to be a member of the relevant professional association in most professions. In industry, especially in industries like the software

development, it is not mandatory to be a member of a professional association. But it is always beneficial to be a member of the relevant professional association. To motivate their employees most professional organizations, provide some sort of incentive to become a member of the relevant professional association. The incentive could be reimbursement of the membership fee or a part thereof, or sponsoring to the seminars conducted by the professional association or some such other incentive. This will enable the employees to gain access to knowledge that is unavailable to them otherwise. Just imagine, you have a troubling issue at work and you meet a number of professionals in a program of the professional association where free exchange of information is in vogue, and an informal discussion with a co-professional would provide a solution to your puzzling problem. The possibility of informal interaction with a number of co-professionals on professional matters is itself a big reward to become members of a professional association and organizations would be wise to motivate their employees to become members of the professional association. The benefits are intangible but are immense to both the individual and the organization.

13. **In-company discussion/bulletin boards** – Some great person sometime back is supposed to have said, thus, "the best helping hand you will ever find is the one at the end of your shirt sleeve". In other words, we need to depend on our resources. Now almost all organizations have some sort of email server in their organizations to serve emails in and out of the organization. The email servers do provide a facility of a bulletin board in which the employees can discuss any issue. Normally these discussions would be restricted to issues relevant to the organization. While we have many employees working on similar assignments, we will not know fully all the talents and skills of our employees. So, we provide these bulletin boards on which a person can post an issue and all those that have a clue about the issue can post his/her view or the solution to the problem at hand or a pointer to arrive at the solution. This facility can also be used to crowdsource viewpoints on an initiative being contemplated by the organizational management. The advantage with these discussion boards is that the participants need not be at one location. They also need not participate at the same time. They need not respond to each post. Each one can read and respond to only those issues that concern him/her at his/her convenient time. However, each discussion needs to be moderated by some one with relevant knowledge. Otherwise, the discussion may deviate from the topic and go haywire.

Again, the employees need to understand that this is not a full-time activity and that they should not stop their regular work to attend to the query posted on the discussion board. The participation in discussion shall be restricted to one's spare time, if it is available. This method can provide amazing results in skill retention if it is administered carefully. As the discussions remain on the board, an individual can search the board to see if somebody had already the posted the issue and received solutions.

14. **Public discussion boards** – Public discussion boards or groups as they are called on websites like Google, LinkedIn and so on, are similar to in-house discussion boards, except that they are hosted on a public server. Issues can be posted and responses to posed queries can also be posted. In these boards, we can tap the expertise of those that are without our organization. The experts need not be not only outside of our organization, but also, they can be outside our city or even country. If we join the right group, we can have access to the best experts in the industry. There are a number of groups in the public domain on various servers available free of cost to us. But we need to camouflage our query so that the public at large would not know that there is a knowledge gap in our organization. If we are not careful in posting our issues, we may expose the lack of knowledge in our organization and this is a big risk and can damage our reputation. So, posting issues on public discussion boards is generally restricted. We normally use a moniker and use hypothetical situation to describe our issue in such a way the reader would not be able decipher from which organization the issues are coming. The great advantage with these boards is that we can access the knowledge of experts from far off distances including those from overseas. The downside of this method is that we can expose the lack of expertise in our organization to outsiders including our competitors.

15. **Mentoring** – Mentoring is quite akin to on-the-job training. It is long drawn out and without a set syllabus and a timeline. Mentoring may be abandoned midway if the candidate is perceived unsuitable. Mentoring is usually carried out by a boss to prepare a subordinate for the next higher responsible position. But of late, the scenario changed. Now, not just managers but even executives who rose up two rungs from the bottom of the executive ladder are being mentored. Now, it is not only the bosses who mentor but other seniors too are becoming mentors of executives working under other managers. Some progressive organizations make it a practice of putting every executive under a mentor. Sometimes, the executive also can select a mentor under whom the executive works.

In some organizations, the details of the executive and his/her mentor would be recorded and at the times of pay hikes and promotions, the views of the mentor also would be taken into consideration along with those of the direct boss. Mentoring is a slow process in which the senior person guides the junior in an informal manner to surmount or circumvent issues at work to become a better supervisor as well as a better person. In the boss-subordinate relationship, there would be some amount of stiffness and unwillingness to share sensitive information or seek certain embarrassing clarifications. With a mentor it becomes easier to share sensitive information or seek clarifications because the relationship is informal. The mentor is like a senior relative like an uncle or an aunt with whom the person shares an affectionate relationship. Most organizations that have this practice of utilizing mentors have found this to be very effective and it helps in retention of talent within the organization. Especially when there is a developing friction between the boss and the subordinate, a mentor can play a crucial role in shaping up the outlook of the subordinate and help the situation. Mentoring in this scenario is mentoring the employee to work effectively, survive and grow in the organization to become a reliable asset to the organization. This practice has taken root in larger organizations but the medium sized enterprises have yet to catch up with this wonderful concept.

The skill development vehicles are summarized in Table 11.1.

Table 11.1 Skill development vehicles

Vehicle	Retention/ Development	Group/ Individual	When to use
Self-study	Retention	Individual	Only 1 or 2 to be trained
Web-based training	Retention	Group & individual	When participants are geographically distributed
Video-based training	Retention	Group & individual	When the participants cannot spare at a stretch
Knowledge sharing	Retention	Group	For problem solving and to share knowledge among colleagues
In-company training programs	Retention & development	Group	When participants are in adequate numbers to justify the cost of training. Most effective of all

(Continued)

Table 11.1   (Continued)

Vehicle	Retention/ Development	Group/ Individual	When to use
Public training programs	Retention & development	Group	Few participants and their time can be spared
Seminars	Development	Individual	To have a glimpse of the new developments for futuristic use
On-the-job training	Retention	Individual	Participants are few and need to be hand held during initial days
Subscription to journals	Development	Individual	For futuristic use
Snippets circulation	Development	Individual	To give info on glimpses of developments with pointers to detailed info
Podcasts	Retention & development	Individual	Any number of participants for self-development during spare time
Membership to professional associations	Retention & development	Individual	Continuous learning for self-development during spare time
In-company discussion boards	Retention	Group	For problem solving and issue resolution to tap expertise available internally
Public discussion boards	Retention	Group	For problem solving and issue resolution to tap the expertise available externally
Mentoring	Development	Individual	For developing individuals to prepare them to shoulder responsibilities of the next higher level

11.5 Components of a Skill Retention/Development Program

Whether it is for skill retention or skill development, it is a program. A program, for our context, is a series of events conducted chronologically on a set of contiguous dates. While there can be larger programs spanning over months, in our context it would be limited for a few contiguous days. We need to arrange for a faculty, a classroom, training facilities like the projection system, coordination costs, administration costs and other incidental expenses like the lunch and snacks and so on. Here are the components of a training program:

1. Course content
2. Training material
3. Course material
4. Faculty
5. Delivery method
6. Evaluation of the effectiveness of the program.

Course content – Course content includes the topics to be covered in the course. We have to be careful while selecting the topics for inclusion. The topics selected ought to be relevant to the skill that needs to be imparted to the participants. Here we need to acknowledge one aspect and that is "knowledge is a continuum that keeps growing every day without any let up". There are many researchers all over the world conducting research on every conceivable topic under the sun adding to the available body of knowledge. Research is being carried out in science, engineering and social sciences. While designing the course content, the one question that is faced by the designers is where to begin and where to end a topic. We need to assume a base level of knowledge in the participants. Then based on the need of the organization, the ceiling of the knowledge needs to be decided. There is a tendency to over-include the sub-topics in the course. Of course, the argument that a little extra knowledge does no harm is OK but the question is "how much extra". It is better to add a little extra subject than to cut down the content. But we should not be carried away in the matter of increasing the content. It is a matter of striking balance and designers have to achieve this carefully. Finally, we need to finally prepare the course content which is an enumeration of the topics and sub-topics to be included in the course. Then we need to schedule the course by allocating dates to the topics and delivery times to the sub-topics.

Training material – While course content enumerates the topics, the training materials contain the details of the topics to be covered. Training material consists of the slides to be used during the program and the notes for the faculty. The slides are used to display the contents on a screen using a PC projection system. Each slide contains bullet points that give a hint to the faculty to elaborate the point further and explain the bullet point. Each slide can contain one or two points with their sub-points. If there are no sub-points in the main points, each slide may contain 3 to 5 points. While the actuals may differ from organization to organization and faculty to faculty, a slide is usually allocated about 5 minutes. So, for a one-hour program, we need to prepare 10 to 15 slides. The faculty notes contain information for the prospective faculty member that would aid him/her in effectively delivering the course. It may contain pointers to books or articles based on which the slides are prepared for further reference. It would also contain quizzes or test scripts for evaluating the course reception of the participants. Training materials are basically aimed at aiding the faculty member in effectively delivering the course and ensuring that the course is easily absorbed by the participants.

Course material – Course material contains the handouts given to the participants during the training program. Two methods are followed in the matter of course material. One way is to give the printouts of the slides used by the faculty to the participants before the course begins. These printouts would have space for the participants to take notes of the faculty lecture against each slide. The advantage is that when the participants look at that slide and the notes taken by them, they would be able to recall the subject delivered. It would act as a memory tickler triggering quick recall. The second method is to provide full text explanations of the topics covered. These texts may specially be prepared by the faculty members or articles which are in public domain or books containing relevant subject matter. The argument in this case is that the participants can take the subject covered in the training program as the base and build their knowledge further by reading these full-length articles or books. As slides are prepared by us, we would have the IPR (Intellectual Property Rights) over them and we can make copies and circulate them. In case of articles in the public domain, we need to check the IPR before making multiple copies. In case of books, we need to purchase adequate number of copies and it can be a costly matter. Most books would place a restriction on making copies of their contents. Alternatively, some publishers sell individual chapters at a much-reduced prices and we can take advantage of that. Some publishers sell electronic versions of their books at affordable prices and they can be kept in the organizational library and can be accessed from there by all the participants. We may select a method suitable for our needs and provide them to our participants.

Faculty – For these programs we can use an in-house faculty or hire an external expert in the subject. Normally, organizations would not employ people just to do teaching. If we do, the faculty would be idle most of the time unless we conduct training programs without any break. Of course, if we are a professional training organization, we would be employing fulltime faculty members but not otherwise. When we conduct training programs, we call upon employees that have expertise in the subject to teach the participants. They also need time off from their regular assignment to conduct the training program. One thing to note is that those who are experts need not be good teachers. So, if we bring in an internal expert to impart a subject, it may not serve the purpose if the selected individual is not good at teaching. External experts are those that have the expertise in the subject but not employed in our organization. We can draw them from the academia, professional consulting organizations, or those that are employed elsewhere but are willing to impart the needed training taking time off from whatever they are doing. We need to compensate them for their time spent with us. Usually, people from academia would be preferred as they are conducting research and have expertise in the subject as well as they are experienced in teaching. Sometimes, we may need multiple faculty members for these training programs. In such cases we may need to use a mix of in-house and external experts as faculty members for the program. When we use internal faculty, we need to allow them time for preparation including refreshing their knowledge, to prepare the needed slides for conducting the training program, to prepare tests and quizzes to evaluate the participant reception of the skill imparted and if it is felt necessary, to prepare course material hand-outs to the participants. This takes time. Therefore, using internal faculty needs more notice period to conduct a training program. As we do not conduct the same training program frequently, the cost of preparing the slides, course material, tests and quizzes would have to be absorbed by only one program. When we invite an external faculty, only a fraction of these costs would be levied on us as the external faculty would recover these costs over a number of training programs. Therefore, it would be advantageous in terms of cost to invite external faculty than use internal faculty. But there would be some situations in which, we cannot or ought not to use external faculty. When the subject is the state of the art or the subject contains proprietary information restricted to within our organization, we ought not to use an external faculty as that would be divulging our proprietary information to them. If we are awarding the work of conducting the training program to a professional training organization, we need not be concerned with the selection of the faculty but need to ensure that the deputed faculty is appropriate for the course. While selecting a faculty, we need to consider these aspects:

1. The individual needs to be expert in the subject at hand
2. The individual ought to possess good communication skills so that the subject can be conveyed properly such that the participants would be able to understand and assimilate the knowledge

3. The individual needs to be able to present him/her well in such a way that the participants can identify themselves with the faculty
4. Preferably, the individual would have gone through some sort of faculty development program sometime earlier.

Delivery method – Delivering the course content to the participants can be achieved in multiple ways. All skill retention/development programs are basically adult education. It is like teaching swimming to the fish! All participants are already working in the field and have a fairly good idea of the topic at hand. The participants would have a clear-cut objective. Therefore, we need to select the right method of imparting the course. The delivery is the most important action in the skill retention/development activity chain. If this is not rendered correctly, all the other activities go waste. We have to be careful in selecting the right method and implement it. Here are the usual methods:

1. **Classroom lecture** – This is perhaps, the most common method. If we use the classical blackboard-lecture method, it would be very dry and would not achieve the objective. Of the five senses of the human being, the sight and the hearing dominate other senses in a classroom situation. We need to engage both the senses. The inputs received through these senses would be processed by the brain at the backend. So, we need to ensure that these three faculties are engaged in a harmonious manner. The input received through the eye needs to be in synchrony with what is being heard and both should make sense. That is why we are using the teaching aids. To engage the visual faculty of the participant, we use the slides projected on to a screen by the PC projection system. We also show items that are pertinent to the topic to further engage the visual faculty of the participants. We compliment the information presented on the screen by our lecture explaining the bullet points presented on the screen. If both are delivered well, then they make sense to the participant and the learning would be maximized.
2. **Demonstration** – Demonstration is another excellent method to impart information to the participants. We actually do some actual work in the organization and do it in front of the participants so that they can see and learn. This would engage all the three faculties of the participants, maximizing their learning. We cannot always utilize this mechanism as most of the work in the service organizations, such as the software development is knowledge-based. Still, we can take advantage of this technique in such activities as systems administration, troubleshooting and so on.
3. **Hands-on training** – This is very useful. It not only engages the three faculties of sight, hearing and thinking but it would also engage the

person in doing. It upholds the adage: I hear I forget; I see I remember; and I do, I know! This would be used to train employees in imparting a new procedure or method or technique. For example, to train programmers in a new programming language, we use this method. We teach some theory and then make the programmers write a program! When we roll out new software for our business operations, we use this method. We explain the software features and make the participants use the software simulating a real-life scenario. The advantage with hands-on training is that, the participants would be fully prepared to implement the training in their work immediately after they complete their training.

4. **Guided learning** – The best example of guided learning is learning how to drive a car. The learner learns a little theory initially and applies it under the watchful eye of the guide. In organizations too, a new entrant to the position would be initially explained a little theory of what is involved and would be working under the close supervision of the supervisor who would act as a guide. There is no set timeline for this except the confidence of the guide. When the guide gains confidence that the understudy has gained adequate expertise that the employee can continue to work effectively on his/her own, then the training ends.

5. **Role play** – Without theory, there can be no expertise. Hence, we supplement the learning of theory with some practical training. But in some situations, such as negotiations and selling, the theory and learning needs to be applied in the field for comprehensive understanding of the topic. Initially, the trainees would be imparted the aspects of negotiation and the techniques to be used in successful negotiation. After learning the theoretical aspects of negotiation, the trainees would be assigned roles and provide guidelines for playing the roles and the goals to be achieved for the role. Then the participants need to conduct themselves adhering to the guidelines specified for the role using their own intellect. There will be no set dialog or strategy specified by the trainer. The trainer would just observe and provide feedback after the role play ended. If the goals are not achieved, then the players failed. When they achieve the goal, their grade would depend on how well they achieved the goal. For example, a trainee playing the role of a salesman is given a goal of allowing a maximum discount of 10% on the price and the trainee ended up giving all the 10% achieved the goal but at the maximum cost. If the salesman allowed only 8% discount saving 2%, he exceeded the expectation. If the salesman failed to clinch the deal, then he/she has obviously failed. The role play, depending on the time available and the importance associated with the training program, would be conducted

multiple times so that the trainees would be expert in the subject at hand. Role play would not only engage the visual, auditory and understanding faculties, it would also engage the thinking faculty of the brain. That way, the learning is maximized.

6. **Motion pictures** – Moving pictures capture a situation which can be exhibited any number of times. To make it easy for you, let us take the example of open-heart surgery. It is not performed every day and it is not possible to allow all the would-be surgeons to witness it in person. If it is filmed or video-taped, it can be shown to any number of participants. Similarly, accident situations, calamities and so on, can be filmed and shown to a number of people to give them knowledge. In organizations too, there are many non-recurring scenarios and we can film them on location or inside a studio with mockups and then use them on our training programs. There are many management and other films available on the market as well as with the consultants which can be rented at an affordable cost to be used in our training programs. Films made leisurely would not only include sights and sounds, they would also include situation-specific background music and the commentary of a narrator explaining the situation and an expert's take on the scene. This would give a long-lasting impression on the minds of trainees which they are unlikely to forget in their life time.

7. **Classroom discussion** – This method uses a discussion between the participants and the faculty member. The participants study the subject on hand and come prepared to the training session. Then the participants and the faculty discuss the subject in depth. In this manner, the faculty member is relieved of imparting the mundane simple aspects of the subject. The participants apply their preparation to their assignments and get their knowledge gaps clarified by the faculty. This concept is already implemented in most of the reputed academic institutions of the world. But it is still to take off in the industry. The argument that the participants are usually busy in their own assignments and find it hard to take time off from their assignments for their preparation has some merit. Hence, the training programs in the industry can be interactive but not a total discussion. If it can be implemented, it would help us to impart much more information to the participants than other modes of delivery.

8. **Group discussions** – Group discussion is akin to brainstorming. The difference is the objective. While the brainstorming technique is aimed at finding a solution to a tough problem, group discussion is aimed at understanding a subject or situation. Group discussions are used

when a suitable faculty to impart the knowledge could not be found. In such cases, the participants prepare for the course as much as they can and then discuss the subject among themselves. Whenever a point is presented well or the group comes to a consensus, all the participants would note that. In this manner, they apply their collective learning and thinking together to improve the knowledge of every participant. In knowledge sharing, one person shares his/her knowledge with the colleagues. In group discussion, there would be no faculty but they could have a moderator who would ensure an orderly discussion but the moderator is also a participant. Group discussions, as a tool for imparting knowledge, throw light on a situation filled with darkness.

Table 11.2 summarizes the delivery methods.

Table 11.2   Summary of course delivery methods

Delivery Method	Faculty	Number of Participants	Merits/Demerits
Classroom lecture	Needed	Multiple	1. Classic and most used method 2. Can deliver most amount of content 3. Can cause fatigue leading to diminished learning
Demonstration	Needed	Multiple	1. Excellent learning 2. Possible only when there is something demonstrable 3. Limited to delivering one or two aspects 4. Not suitable for delivering theoretical aspects
Hands-on training	Needed	Multiple	1. Excellent for imparting theory plus practice 2. Suitable only when hands-on is possible
Guided learning	Part time	1 to a maximum of 5	1. Very effective in developing expertise 2. Not suitable for imparting theoretical knowledge 3. Number of participants is limited
Role play	Needed	Few participants	1. Very effective in imparting knowledge 2. Limited applicability

(Continued)

Table 11.2   (Continued)

Delivery Method	Faculty	Number of Participants	Merits/Demerits
Motion pictures	Not needed	Multiple participants	1. Very effective 2. Production cost is very high
Classroom discussions	Needed	Multiple but not too many	1. Can impart large amounts of information 2. Needs time off to the participants for preparation
Group discussions	Only a moderator	Multiple but not too many	1. Can throw light on a completely new area 2. Needs committed, knowledgeable and prepared participants

11.6 Evaluation of Participant Reception of the Course

Measuring the effectiveness of a training program has been a paradox from the beginning. Measuring the faculty performance is easy; measuring the participant reception is easy; but measuring the impact of the training program is not easy at all! Every management person would ask if the training program justified the Dollars spent on conducting the training program including the cost of the time spared for the participants. There are no real answers for this question. It was granted that the benefits of the training programs are intangible. In the case of the skill retention programs, we measure the impact of the training program after which the employees are able to continue working in the changed environment at the same effectiveness or improved effectiveness. But in the case of skill development programs, it becomes much more difficult to see those benefits. For example, how would we measure the benefits of reimbursing the expenses for membership in a professional association or the cost of subscriptions to professional journals or the money spent on sponsoring our executives to a seminar?

It is not our intention to say that it is futile to attempt measuring the impact of a skill retention/development program. We need to ensure that each cent that is spent in the organization received a concomitant and commensurate benefit for the organization. We ought to measure the impact in whatever way we can. We can easily measure the reception of a program immediately after the program is concluded and we should. As soon as a program is finished, we need to obtain the feedback of the participants. Table 11.3 provides some hints on taking the participant feedback.

Table 11.3 Hints for taking feedback on skill retention/development programs

Delivery Method	Feedback Hints
Classroom lecture	Immediately upon a faculty member concluding the lecture and as part of the program
Demonstration	Immediately upon the conclusion of the demonstration and as part of the program
Hands-on training	Immediately upon the conclusion of the program
Guided learning	Immediately upon the declaration by the guide that the learner is ready for work
Role play	Immediately upon the conclusion and as part of the program
Motion pictures	Immediately upon the end of the film
Classroom discussions	Immediately upon the conclusion and as part of the program
Group discussions	Immediately upon the conclusion and as part of the program

When we take the feedback as soon as the program is concluded, there are two aspects at play: one is that the program details are fresh in the minds of the participants and they can accurately rate the program; and the second aspect is that there is the recency effect. Even if the program did not cover all the details or gave inaccurate information, the participants would not come to know of the defects in the program until they had fact-checked the information received. So, this feedback taken immediately after the program can give erroneous results. Therefore, we need to take another feedback after some time. We may take this second feedback after the lapse of about 3 or 6 months depending on the organizational environment. The period should be so chosen as to give the participants a chance to apply the learning in their work and see the efficacy of the program.

We need to obtain the feedback covering these aspects:

1. Adequacy of the topics included in the program
2. Adequacy of the actual depth of the coverage of the included topics
3. Adequacy of the examples, quizzes, tests used in the program
4. Suggestions for addition/deletion/modifications of the topics included
5. Suggestions for increasing/decreasing the duration of the program
6. Feedback about the hands-on sessions
7. Feedback about the guiding process
8. Feedback about the discussions
9. Any other suggestion for improvement.

Now, we need to tailor the format for each type of skill retention/ development program we conduct. We suggest that it is better to rate the aspects on a 5- or 10-point scale so that we can subject the scores to statistical analysis. As you can see, some of the aspects, specifically about suggestions would be textual form and we need to take this as a feedback and have a mechanism in our process to dovetail these suggestions into out future courses.

Then during the delayed feedback taken after a lapse of 3 or 6 months, we may provide the feedback taken immediately after the program to the participant to allow him/her to consider changing their feedback after putting the learning to work.

It is better to obtain the feedback of the faculty and some organizations have implemented this aspect. Faculty can throw light on the way the participants received the course content. Sometimes, the participants do not evince keen interest in the program and give a negative feedback on the course as a measure of defense of their inability to benefit from the program. This will be revealed by the faculty feedback. Many times, the participants included in the program are not the right ones for the program. In such cases, the program cannot be effective because these disinterested participants get into arguments and cause hindrances in the effective delivery.

We suggest taking faculty feedback on these aspects:

1. Adequacy of the size of the batch
2. Adequacy of the duration of the program
3. Adequacy of the facilities provided for conducting the program
4. Homogeneity of the participant profile
5. The participants that participated the best in the program
6. The participants that did not participate adequately in the program
7. Any feedback received by the faculty from the participants
8. Any suggestions for improvement.

It is not necessary to obtain a follow up feedback from the faculty member as he/she would not be using this course in his/her work later on.

Then we may also obtain a feedback from the supervisors of the participants in the program after the lapse of about 3 to 6 months' time. This will allow them enough time to observe and notice any change in the way the participants conduct themselves at work as well as the changes in the productivity and quality of their subordinates as a result of participating in the program.

We need to obtain their feedback on the following aspects:

1. Did the program have any impact at all on the participant
2. Whether the impact noticed is positive or negative

3. Did the program prove beneficial or waste of time to the organization
4. Would he/she recommend this program for future participants
5. Any suggestions for improvement in the program.

What is often neglected and rarely practiced is the postmortem of the programs that were conducted in the organization. Once all the feedbacks are obtained for a program, we need to conduct a formal postmortem of the program. We may invite concerned senior executives that sponsored this program and the coordinator needs to present the highlights and the lowlights of the program. If we are a process driven organization, we may conduct an audit of the concluded program after all the feedbacks are received and the auditor may present the program to the audience. The audience may give their feedback and all aspects of the program can be discussed in the greatest detail. We can draw suggestions for improvement in the future programs and improve them when we conduct them again in future. It can also improve our competence in conducting skill retention/development programs for the betterment of the organization.

11.7 Quality Assurance in Conducting the Skill Retention/Development Programs

Every activity performed in professional organizations must be subjected to quality assurance. Quality assurance consists of two sets of activities, namely, the defect prevention activities and defect detection activities. Defect prevention activities include defining a process, procedures to perform the activities, standards and guidelines to ensure a minimum level of quality in all the activities performed formats and templates to ensure capturing of information uniformly irrespective whoever is capturing the information and finally checklists to ensure that all activities are carried out comprehensively. These are in the domain of the organization but the individuals working in the HR department have the onus in participating in the process definition and improvement activities.

Quality control activities consist of the activities performed while the work is being carried out to ensure that there are no defects in the deliverables. Usually, these consist of peer reviews and testing if applicable. In skill retention and development activities, testing is in the manner of conducting evaluation of the participant reception of the course. In the courses, the faculty becomes crucial as it is the faculty that delivers the content. The success of the course depends directly on the performance of the faculty. Of course, we train the faculty members in effective delivery of the course but we need to ensure that the faculty uses all those delivery skills in every program. On-the-job audits are a standard practice in many service organizations. Doctors are audited by experts; airline pilots are audited; drivers are audited; all this auditing is carried out when they are performing their operations. Educational institutions also audit their teaching staff periodically while they are teaching. However, this auditing is not carried out every time they are working. This auditing is performed periodically on a defined schedule. Similarly, we also need to audit the training programs while they are being conducted on all aspects. If the audits uncover any NCRs (Non-conformance Reports which contain any deviations from the defined process), we need to take necessary corrective actions immediately to correct the present situation and preventive actions to prevent the recurrence of such defects in future programs.

We also need to conduct periodic audits on the HR department on how the activity of skill retention and development activity is being carried out in the organization and if any deviations from the defined process are found, NCRs need to be raised and closed by taking necessary corrective and preventive actions.

In this manner, we ensure that quality is built into the activities of skill retention and development being performed in the organization to ensure that the organizational human resources are retained at the cutting edge of technology applicable to their respective roles and responsibilities assigned to them by the organization.

12

Attrition Management

12.1 Introduction

People keep leaving organizations. People leave the best of the organizations for various reasons. People get better opportunities elsewhere with higher pay or rank and leave the organization. People leave organizations for personal reasons to be close to their relatives or properties they own or for health reasons to be in a suitable environment or any other such compelling reason. People retire and have to leave the organization. People also lose their health or meet with accidents and sometimes, the sad event of death. Sometimes we send people out of the organization for poor performance. People, for whatever reason, leave the organization.

We need to accept the fact that we lose people but we have a duty to ensure that no organizational operation suffers for lack of adequate number of people with right skills at the right time. Of course, we have our recruitment pipeline that is kept ready to replace the people who leave us but we need to derive whatever benefit we could for our organization even from people leaving us. Let us now discuss all aspects of attrition of people from the organization.

12.2 Reasons for Attrition

It is a fact of organizational life that people leave their organization for various reasons. Primarily, attrition is of two types:

1. Planned attrition
2. Unplanned attrition

Planned attrition refers to those people whose departure from the organization is predictable. Unplanned attrition refers to those people whose departure from the organization is unpredictable. All the reasons enumerated below can be fit into one of these two classes.

1. Resignations – unplanned attrition
2. Incapacitation – mostly unplanned
3. Deaths – unplanned, mostly
4. Retirements – planned
5. Downsizing – planned

When planned attrition takes place, we are ready for such events. But we would not be prepared for unplanned attrition. We are caught off-guard. Of the three 3 varieties of unplanned attrition, attrition to resignation from our organization is the most common and we should learn about it in depth and understand the event to be prepared for it. Let us discuss this inevitable event and understand its repercussions.

Resignations – Why do people resign from their jobs? There are a variety of reasons. Some of these reasons are personal to the employee but some are attributable to the organization. Here are the reasons attributable to individual aspirations:

1. **Better pay** – Most employees, if not everyone, desire more money. It is a common need. But money is not everything in life. Satisfaction in the job and the ability to hold one's head high while talking about one's employment and the work rate higher than the money. For example, being a gangster or an assassin may bring in more money but one cannot be proud of doing it or talk openly about it with satisfaction. Still, money becomes a deciding factor when the jobs are similar in nature. Most resignations happen due to this reason. Unless the employee is crucial to our organization owing to the specialist skills that employee possesses or critical to our organization, we may let that individual go. But if the employee is needed, we may need to match or exceed the expected monetary gains, the employee expects from changing the job.

2. **Better perks** – Most organizations offer some part of the salary as perks. While the money for the perks is taken from the cost-to-company salary, still, they make the individual happy. For example, a chauffeur-driven car for transport is a big ego-booster for senior executives and it has the potential to make a senior executive change the jobs! Better healthcare, subscription to gyms, clubs, reimbursement of expenses for professional associations and journals and so on are some of the perks that wean away employees from our organization. Matching these would require our organization to bring in a policy change because we need to apply those changes to all our employees. When employees leave our organization

for better perks, we can only show other advantages of staying with us and allow them to go if they do not relent.

3. **Better designation** – This is more or less a perk. Especially smaller organizations use this ploy to attract talent from larger organizations. To become a Vice President in a large organization takes years of toil where as it is easy in smaller organizations. When one of our employees is lured away by another organization using a nice sounding designation, perhaps, we may not be able to retain such employees. Of course, if the employee is just one or two levels from the designation offered to him/her by the recruiting company, perhaps we can convince that employee to remain with us and offer accelerated promotions to retain the leaving employee. Otherwise, we need to let that person go.

4. **Higher responsibility** – Some individuals seek higher responsibility rather than pay or designation. They would like to handle more activities of diverse nature or different departments. A delivery head would like to handle quality assurance or marketing in addition to the existing responsibilities or a HR head would like to handle finance in additions to the existing responsibilities and so on. Such people are likely to change organizations if they get the opportunity to handle additional or higher responsibilities. In such cases, it is very difficult to provide such opportunities in our organization itself. Such actions would give rise to stiff resistance from other peers. It is perhaps better to let such people leave the organization.

5. **Bigger organization** – This is a situation frequently faced by smaller organizations. Some people prefer the anonymity offered by large corporations. There is a saying that it is better to be a big fish in a small pond than being a small fish in a large pond and most people adopt this adage. But some people prefer to be small fish in a large pond than being a big fish in a small pond. When employees leave our organization for this reason, we cannot retain them as we cannot grow our organization to be a large one overnight. This is a case opposite to that of seeking higher responsibilities that go from a larger organization to a smaller organization. In such cases, it is better to let them go.

6. **Organization with better reputation** – Organizations that have a high reputation for quality products or services are a magnet to employees and when they get a chance they jump over to those reputed organizations. But, unless their present position is protected, employees would not jump ship. So, when employees resign from our organization to join a company that has better reputation than ours, we can reason with

that individual about the benefits of continuing with our organization. Reputation of the organization does not give any new or additional benefits to the employee. We need to bring this fact to the notice of the employee and contrast our career advancement opportunities and salary policies to try and convince the employee to continue with us. Perhaps, we may give that employee some extra pay or perks or responsibilities and post that person in a department of his/her liking or some such other thing to try and retain that employee with us.

7. **Better location for personal purposes** – People change organizations to be in a better location. The new location could be new city or town which offers them better facilities. Employees may like to improve their educational qualifications by going to evening school and if it is not available near to the location of our organization, they may change the job and the organization so that they can fulfill their aspiration. Similarly, an employee who begets children may leave to another place so that they are close to the school district that has great schools for his/her children. Sometimes, an employee may change his/her job and the organization and move to a location that offers employment opportunities for his/her spouse. It can also happen when the spouse gets a better job and moves to another city as our employee would like to keep the family together. Most organizations have a policy not to employ the spouses and, in such cases, we may perhaps let the employee go. But if our organization does not have such a policy or we have suitable opportunity for the spouse, we may offer employment to the spouse of our employee and retain him/her with us. There could be other reasons to change to a better location and people leave our organization.

8. **Closer to one's properties** – Rare such instances may be but, in some cases, people leave our organization just to be close to their ancestral properties. If one of our employees inherits some valuable property in another town, such employees may wish to be closer to those inherited properties and leave our organization. Of course, we now have specialized agencies to effectively handle property management and we can perhaps assist the employee to find an appropriate property management agency and try to retain that individual.

9. **Location with more congenial weather** – Some people have weather problems. Some suffer heavily in cold weather while some others suffer from hot weather. In countries like the USA which has locations with varying weather, it can happen that a person moves from a hot weather town to a cold weather town and falls sick in the next few years when that

individual's immunity system wears down. Medical treatment would not help such individuals and only a change of place to a location with hot weather would help such individuals. Sometimes, females with ancestral hot weather locations may not be able to conceive in cold weather places. They may need to move to a location that is less cold and hotter than the present location. Asthma is another ailment that needs hotter weather. There could be other such medical reasons and we need to let them go because we are incapable of changing the weather of the city in which our organization is located!

10. **Stagnation at present position** – This usually happens in large organizations and sometimes in medium sized organizations too. If some one is near the top and cannot be promoted to the next level because that position is occupied by some one who has to be either promoted or transferred to make it vacant. Sometimes, qualification or skill limitations make it impossible to promote the employee to the next level. Or the department in which the employee works reached saturation and stabilized, there may not be room for promotions. In such cases when an employee stagnates in the same position for longer durations, that employee may seek outside opportunities and leave us if a suitable offer is received. In such cases, we just need to allow the employee to leave us.

Now, let us look at reasons for attrition that are attributable to organization.

1. **Better opportunity** – In certain specializations, people jump ship for better opportunities. For example, in research specialization, people change organizations if they get an opportunity to work on a ground-breaking research opportunity or an opportunity that may have the potential to win a Nobel Prize. In some cases, if there is an opportunity to work on a project that serves the people at large or something like that. Such opportunities keep coming up and they are competed for in a stiff manner. Still, people compete and if they win the competition, they would jump the ship. Some academicians would like to work in industry and the people from the industry would like to work in the academia. In some cases, a head of the department may get an opportunity to head an entire organization. In such cases, it is well nigh impossible to retain a person pursuing his/her passion. It is better to let them go.

2. **Working conditions and ergonomics** – Working conditions are very important for people. Working conditions include the processes, procedures, administrative mechanism, tools and workstations. We can go

overboard in the matter of processes and procedures in our eagerness to ensure that our deliverables have built-in quality. Similarly, the administrative mechanisms become too daunting for the employees to resolve their issues. The administrative staff, in their eagerness to safeguard and protect the organizational interests may carry it too far and treat employees unfairly. Professor Herzberg included these working conditions and ergonomics in the hygiene factors in his two-factor theory of motivation. We have a need to maintain good working conditions which are congenial to the employees. We need to periodically look at our processes, procedures and administrative mechanisms to see that they are employee-friendly. It is easy and possible to modify these without major investment or inconvenience. However, when it comes to workplace ergonomics, we design our workplace and workstations with the contemporary technology and equipment but as the time passes, newer technologies and equipment become available and newer organizations design their facility with that newer technology. The technology is getting upgraded once every three to five years. In this age of fast obsolescence, can we change the equipment and technology of our facility every five years? We would say, very difficult if not downright impossible. However, we need to ensure that the ergonomics of our facility are maintained at such a level that they do not pose a health hazard to our employees. We need to upgrade the ergonomics of our workplace as necessary. We also need to ensure that the general environmental conditions of our workplace to be as pleasant as possible, even if we do not upgrade our facility with the latest equipment. We need to periodically revisit our processes, procedures, and administrative mechanisms to ensure that they are employee friendly. In this manner, we can retain employees who would otherwise leave our organization in search of better working conditions.

3. **Supervision** – Supervision is one of the major causes of employee attrition. We might have conducted training programs before we promote people to supervisory positions but when we recruit persons from outside our organization into supervisory positions, they may not be as well trained as our people are. Left to themselves, supervisors have a tendency to treat their jobs predominantly as policing! When this policing becomes prevalent, the people, especially in the category of professional workers, working under such supervision are likely to be peeved to the extent that they may resign from our organization. We need to ensure that our employees in supervisory roles, have an understanding

of this phenomenon and take care that they do not exhibit the policing personality to their subordinates. Attrition for this reason should not be allowed to take place in our organization. When the first instance of such occurrence comes to our notice, we should take immediate action to correct this situation before the trickle becomes a deluge. We should retain the employee who resigned due to this reason by changing his/her department or some such other action agreeable to the leaving employee. Then we need to correct the situation. We also need to investigate why our periodic audits did not discover this practice and plug in all loopholes to prevent its recurrence.

4. **Lack of career advancement opportunities** – This happens in mature and saturated organizations. While the job may be secure, the opportunities for career advance would be scarce. In such cases, employees may leave preferring growth over stability. In some rare cases, especially in large organizations there may not be many career advancement opportunities. It happens in the case of individuals recruited from less reputed educational institutions who would be placed in slow track in the matter of career advancement. Such people would hit the career advancement ceiling at some specific position beyond which that individual would not be promoted. In some cases, some people even with Ivy League pedigree would hit the career ceiling if their performance was not up to the expectations of the management. Large organizations would, sometimes, not like to let such people go because it may send a message that the organization is firing people even from Ivy League pedigree which may propel other people to resign. In such large organizations, employees recruited from highly reputed educational institutions (generally referred to as Ivy League pedigree) would be placed in fast track career advancement. Such organizations do have unwritten policies that would allow only people from Ivy League institutions to occupy the positions of heads of department and above. Now, the only way a person, not from the Ivy League institutions, can advance beyond a certain level is to go to another organization, perhaps smaller in size and reputation to break the career ceiling. Of course, exceptions are always part of the organizational life where persons without the Ivy League pedigree can break the ceiling but it is rather the exception than the rule. When such people leave the organization, they are usually allowed to go and they would be sent off with affection and excellent referrals. That is the best option in our opinion too. If we place a ceiling on the career advancement on some individual, is it not better to allow him/her to go?

5. **Stability of employment** – Most organizations follow a policy of ensuring a long tenure for their employees. Some organizations do not follow this policy. Ensuring a long tenure for the employees was easy in the bygone days. Then people were willing to work without much career advancement doing the same type of work year after year. Of course, that was in blue collar category. In the new category of professional workers, people are not willing to do the same thing year after year for a number of years. The attitudes have gone through a metamorphosis, not only for the employees but also for the employers! The organizations want their organizations to be lean and mean. That translates to minimization of costs of which employee costs form a major part. The organizations, especially, in the services sector, have adapted a flat hierarchy than the classical pyramidal structure of the yore. There were more than 8 levels of management which came down to about 5! With the structures becoming flatter, there are fewer management levels that can be filled with employees who have put in 5 years of tenure. Some people, who joined at the entry level five or more years ago, need to go! So, managements willfully create an atmosphere that encourages employees to leave the organizations on their own. So, the employees knowing that the management is following this type of policy, naturally wish to leave at the earliest opportunity and the management gladly releases such resigning employees. It is a made-for-each other scenario. We can't improve upon it!

6. **Retirement benefits** – Organizations differ from each other in the matter of retirement benefits. Of course, the days of retiring in the same organization in which the employee took up his/her first employment has come down drastically. Very few employees, if at all, retire in the same organization in which they began their career in the present day. Employees are changing frequently, so frequently that they do not put in long enough tenure to attract the retirement benefits. All the same, this is a factor for employees to compare while changing the organization, especially for employees above fifty years of age when they like to plan for retirement. Some of the retirement benefits offered by organizations are mandated by the government regulations but some are not. Those organizations that offer better retirement benefits have a better chance of retaining their employees better than those that offer just the mandated retirement benefits. Of course, the total salary package can offset the better retirement benefits. So, if we are an organization that offers better retirement benefits than comparable organizations, we have an edge over

our competitors. But if we are not better than comparable benefits in the matter of retirement benefits, then we may not be able to retain our employees by better total salary package. Otherwise, we need to allow those employees to go when they choose.

7. **Lack of opportunities to showcase abilities/creativity** – Especially in the case of professionals like the programmers, financial consultants, lawyers, scientists and the like, they generally have a yearning to showcase their talent and get recognition. If they were assigned routine mundane tasks, they can feel frustrated and leave our organization. We know that there are mundane tasks that need completion and that need to be executed by these professional workers. We can also assert confidently that professional workers acknowledge this fact. What they cannot digest is the fact that mundane tasks are continuously assigned to the same persons and tasks with the potential to get recognition are assigned to the same blue-eyed boys/girls again and again. Work is the greatest motivator as we discussed in Chapter 9 on motivation and morale. If the opportunity to showcase talent is denied, the professional workers leave us. What we can do is to adopt a transparent system of work allocation that would rotate the mundane tasks and creative tasks between different individuals in which every one gets an opportunity to work on tasks that have high visibility and the potential for recognition. When an individual wants to leave our organization, we need to immediately vet our work allocation system and modify it if necessary, so that all people would get an opportunity to work on challenging assignments. We can convince the leaving employee that he/she would get a fair opportunity, henceforth, and retain that leaving employee. We have to allow that person to leave if the changed system (or proposed changes in the system) does not convince that individual.

8. **Perceived injustice** – This thing happens, especially when promotions to key positions are carried out. We fill our senior positions through internal promotions and external recruitment. In either case, it is possible that there are some employees in our organization, who feel that this position should have been theirs. They perceive injustice in either promoting or recruiting someone else. Organizations define their policies with the twin objectives of motivating the employees towards better performance and at the same time, protect the interests of the organization. One cannot be sacrificed for the sake of the other. It is a fine balancing act like the tightrope walk. In this balancing, especially when these policies are changed to stay current with the changing statues and environment,

we may introduce some policies that are perceived as detrimental to the interests of a section of our employees. For example, we sometimes, place ceilings on the level to which employees, with lower educational qualifications, can be promoted. This would affect the interests of that section of employees and it would be heartburn, if the employee is at the ceiling when the new policy is promulgated. Can we really do anything about it? Not really. But sometimes, exceptions are given especially to those employees who are the ceiling level when the policy is promulgated but that would cause heartburn to those employees who are at just one level below the ceiling. Some organizations would not restrict the promotions to less qualified individuals but place an efficiency-bar which involves conducting some sort of tests passing which will make the individuals to cross the bar and become eligible for the next level. Some organizations would exempt the people already employed in the organization from the ceilings but would apply it to the people joining later on. What if a person joins our organization the next day of promulgating this ceiling policy? When that individual attended the interview, this was not revealed to him/her because this policy was yet to be promulgated. There would be many situations like this especially when policies are changed. So, it is possible that some individuals feel that injustice was done to them. And when such people leave, it is incumbent on us to allow that person to go but we should ensure that such departures are carried out smoothly and with utmost respect to the individual.

9. **Lack of opportunities for change of specialty** – This is a common occurrence in organizations. For example, a programmer who was using COBOL would like to changeover to Java programming and looks for a chance to change his/her specialty. A legal person would like change his specialty from corporate law to civil law. A scientist may like to change from asthma research to cancer research. A nurse would like change from ward duties to operation theatre duty. There can be many such other aspirations in the individuals working in our organization. We at the HR department collect these aspirations during the performance appraisals and try to accommodate these requests. But it is always not possible. We would go out on a limb and say, that it is not possible to accommodate these requests most of the times. When the team is well set, we, as also the departmental head, would not like to disturb the equilibrium. No one likes to disturb a successful/winning team just for the sake of the whim of an employee. Besides, as some would argue, the employee signed to work in any department at the exigency of the organization. Most

employees would understand and cooperate. But their desire can keep burning inside them and they seek that opportunity elsewhere and we may lose an excellent employee otherwise. When employees come up with such requests which they feel is essential for them to focus one 100% on the work, we need to do something. We should take steps to seek such other opportunities within our organization and this must be transparent and visible to the employee. We at HR need to coordinate this process and help the employee to achieve his/her personal goals along with those of the organizational goals.

We at HR department have a major role in the people leaving due to reasons attributable to either the individual or the organization. Our efforts ought to retain all good employees and work toward that objective. Let us now discuss the modes of attrition.

12.3 Modes of Attrition

Employee attrition takes place in different modes. There are two possible modes of attrition, namely, the employee-initiated attrition and the organization-initiated attrition. Let us now look at both these types of attritions.

Employee-initiated attrition – This attrition is initiated by the employee. The following are the modes of employee-initiated attrition:

1. **Clean separation** – A clean resignation involves the following:

    (a) Giving the full and required notice conforming to the rules of the organization – Different organizations would have different notice periods stipulated for the employees who desire to leave the organization at their own will. It is usually two weeks but it could be different in different organizations. When crucial resources like the heads of department or the senior management personnel resign, they may be required to stay a little longer to transition the responsibility to the next executive. So, a clean resignation involves giving the stipulated notice for quitting on friendly terms with the organization. An individual using the clean resignation route would stay the full notice period and offers full cooperation so that the organizational functioning or financial stability is not adversely affected. Organizations generally show some soft corner for such employees and if a future opportunity arises, they may re-employ such individuals once again.

(b) Clean handing over of charge to the person designated by the supervisor. Whenever any employee leaves the organization, there would be quite a few activities being performed by that person and quite a few plans would be dependent on that individual. So, it is incumbent on the leaving individual to handover all the pending activities to the designated person. It may sometimes include training that designated person in performing the activities. Sometimes it may involve introducing the designated person to the customers, vendors, consultants or such other persons so that they would not have any confusion about whom to interact with after the resigning executive stops handling the activities. Sometimes, it may involve handholding the designated person for sometime to make that individual comfortable in ensuring that the organizational activities do not suffer for even a minute due to an employee leaving the organization.

(c) Returning all the assets held by the resigning employee to the designated agencies. All employees hold some assets for official use during their tenure in the organization. The assets include the workstation, the hardware and software tools, information artifacts like books, documents and program code, and such other assets. These need to be handed over to the concerned organizational agencies and obtain a no-dues certificate from each of those agencies. A clean separation involves that all the organizational assets are handed over in proper working condition and defects if any, would be pointed to the receiving agencies.

(d) Pending work – Whenever an employee submits the resignation, there certainly would be some work that is pending to be completed. While handing over the pending work to the designated person, the resigning employee needs to continue functioning till the day of physically leaving the organization. Of course, the designated person needs to be eased into the position. Still, the resigning person needs to function until the designated person is ready to take over and then handhold the person as long as the employee is not relieved from the organization. The resigning employee ought to ensure that no organizational activity is delayed, or is affected in terms of quality. There may be some task that only the leaving employee can complete and such tasks need to be completed before relinquishing charge.

(e) Final leaving – The final leaving action involves handing over the identity card, office cell phone and such other assets which are held until the last day, need to be handed over to the designated agency which is usually the security department. The resigning employee ought to obtain a no-dues certificate and hand it over to the designated department which is usually the HR department which ensures that all dues to the employee are paid on time. The employee also ought to obtain the relieving certificate from the HR department. When an employee leaves, there ought to be nothing pending from the employee to the organization.

2. **Dirty separation** – Dirty separation would not follow any procedure. Still these are the characteristics of dirty separation:

(a) Usually, there would be no proper resignation. If it is there, it may not give the stipulated notice. The resignation letter may be given on the day before leaving or received by the organization after the employee stops coming to the work.

(b) The handing over of the charge may not happen or happen in a haphazard manner. The next person would not at all be trained or trained in a lackadaisical manner. The organizational activities are likely to suffer because the leaving employee did not hand over the charge properly.

(c) The organizational assets would not be handed over properly. They may hand over some assets but may not hand over completely. If the assets were handed over completely, they may not be in proper working condition and the defects therein were not pointed to the receiving agencies. Sometimes, the leaving employee may not obtain any no-dues certificate from these receiving agencies.

(d) Pending work may be left midway while leaving the organization. The designated person may not be adequately trained by the leaving person. No handholding would be performed by the leaving employee. Organizational activities are likely to suffer.

(e) Final leaving may not be smooth. The individual may not wait to receive the relieving certificate. Sometimes, the HR department may not be in a position to issue the relieving certificate because the leaving employee did not obtain and submitted the no-dues certificate from the concerned agencies. In such cases, the HR department needs to coordinate with the concerned agencies to obtain the dues of the employee so that they can be recovered from the payments due to the employee.

3. **Fleeing** – Fleeing is leaving the organization without submitting the resignation or giving any information to anyone in the organization – It usually happens the day after the pay day. No notice whatsoever would be given even to the immediate supervisor. The individual just becomes absent and it takes a couple of days to notice that there is something fishy in the absence. Enquiries with colleagues and other known friends would not reveal any untoward incident like an accident or illness. The individual also would not be traceable. The phone may not be reachable or not answered. This happens especially in the case of people who are under some sort of contract with the organization with stipulations for penalties for not completing an assignment or working in the organization for a specified length of time. Sometimes, people may become involved in some sort of crime which causes the individual to abscond not only from our organization but also from the known address and without information to anyone. When a person just becomes absent, how should we handle the situation? Initially, it would be the supervisor directly supervising the work of the absconding individual that comes to know about the absence of the employee. But we need to give benefit of doubt to the individual because he/she might have met with an accident far away from the usual place of residence and lying unconscious in a hospital bed where no one recognizes the person. Organizations generally have a policy for handling unauthorized absences like this which stipulates a certain period like a week or two weeks to wait until the organization begins to treat such absence as unilateral separation by the individual. Once this grace period expires, the supervisor initiates the separation activities, the first of which is to inform the HR department about the unauthorized absence. Then, the HR department along with the supervisor performs the release activities, which will be dealt with in the subsequent section.

Now that we discussed the modes of employee-initiated attrition, let us now discuss the organization-initiated attrition.

12.4 Modes of Organization-Initiated Attrition

Organization initiates employee separation:

1. Retirement – every organization would have some specified age to retire employees. When a person completes that age, the person needs to retire from work. The retirement age varies from organization to organization

and from position to position. The senior management personnel would have a higher age limit for retirement than the persons at the working level. If the person is using the mental faculties, then retirement age would be higher. An individual who is working as a designer learns so much over the years and would have immense knowledge which cannot be sacrificed just because that individual attained some age. Similarly, a marketing manager gathers so much market intelligence and contacts and hence cannot be allowed to go because all his/her contacts would walk out of our door along with that individual. Therefore, the retirement age is flexible. But when a person is retired at his/her own volition or at the decision of the organization, that separation is initiated by the organization.

2. Termination/dismissal – When an individual is working in the organization and that person's performance is not satisfactory, we spend efforts and time to improve the performance through improvement plans, job rotation, special incentives and so on. But when all those efforts fail, we have no other recourse but to let that person go. Usually, most organizations allow such person to submit a resignation and leave as that would allow the organization to put the reason for leaving as "resignation" rather than as "termination/dismissal". While the words "termination" and "dismissal" have subtle difference, the end result is the same, that is, the individual is sent out of the organization. The term "termination" connotes that while the individual committed an unpardonable act, it was not considered so big that an opportunity cannot be given to the individual for explanation. That individual is separated from the organization only after an enquiry confirms that the individual is culpable and that the act deserves separation. The term "dismissal" connotes that the person was caught red-handed and that the person cannot be allowed to continue in the organization any more. There was no doubt about the culpability of the person. Every professional organization would have lists of acts that attract termination and dismissal. Based on this list and the situation surrounding the perpetration of the act, the decision whether to terminate or dismiss would be taken. When a person is terminated, the employment prospects are brighter than when the person is dismissed. In these cases, the organization would initiate the separation.

3. Death/disability – These two are unfortunate events. But as death is an irrevocable event, we cannot continue the person consumed by death on our organizational rolls. We need to initiate his/her separation.

Disabilities are of many varieties. If an employee working in our organization meets with an unfortunate accident which results in a disability for the person, we need to evaluate that person's ability to continue in employment, if not in the same position that individual held before the accident, in a lower or other position in which the disability does not hamper the work. Only when we are certain that the disability is such that the person cannot continue employment in our organization in any position, we need to initiate separation.

4. Downsizing – Technologies become obsolete and new technologies keep coming into the market. New organizational competitors come into the market and chisel away at our market share by using the latest technology and tools that reduce their costs. Therefore, organizations periodically carry out an exercise in re-engineering our business processes and reduce the headcount to stay competitive in the market and take advantage of the newer tools and technologies. When this re-engineering happens, we need to identify those people who cannot be re-trained and posted to other useful activities. And then we need to let them go with full respect and an excellent reference letter. In such cases, organizations initiate separation.

Now having discussed the modes of attrition, let us now turn our attention to release activities.

12.5 Release Prerequisites

The final action in the attrition chain is the release activity in which the employee is released from the organization. Once released, the employee becomes an ex-employee and would not be able to come inside the organization except with a visitor pass! The first release activity that we need to perform is to derive as much benefit for the organization as possible. True, the separation of an employee is a loss-making proposition as we need to incur the costs of recruitment and training the new recruit. But, even from such situation, we derive the benefit of learning useful lessons that are otherwise not available. Employees hesitate from giving frank opinions about people, policies and events for the fear of backlash while they are still in employment. When they leave, they are less likely to be apprehensive and give their frank opinions and that can benefit our organization. For this purpose, we conduct exit interviews with the leaving persons. Let us now discuss the exit interviews.

Exit Interviews – Exit interviews are an excellent opportunity for us to get a frank feedback from one of our employees. We get honest feedback because the employee does not fear any backlash as he/she will be leaving the organization shortly. When should we conduct this exit interview – immediately after submitting the resignation or just before giving the relieving letter? While many organizations follow the practice of conducting the exit interview as a prelude to issuing the relieving letter, in our humble opinion, it is better to conduct this interview soon after the individual submits the resignation. Of course, the immediate supervisor performs this activity and records it and passes it on to HR through proper channel along with the resignation letter. But that would be the reason given in a defensive manner so as not to hurt the feelings of the immediate bosses. If the trained HR persons conduct the interview in a careful manner, a host of information is likely to come out from the leaving person. That individual can offer information about the inefficiencies, instances of favoritism, any perceived injustices, biases/prejudices existing in the workplace, existence of informal groups at the workplace, suggestions for improvements and so on. All this information needs to be captured. This information can be fact checked for accuracy before the employee leaves the organization and verify it back from that individual for keeping the accurate information or correcting our own impressions. We can also use this opportunity to try and retain the employee by finding out the real reason for leaving and see if that can be resolved to the satisfaction of the employee. Exit interview needs to be formally conducted by a person trained in conducting such interviews but in such a manner that the leaving person feels confident to reveal sensitive information. The minutes have to be captured and made part of the employee record. During the exit interview we also need to elicit the willingness of the employee to come back to our organization when a future opportunity arises that is mutually acceptable to both our organization as well as to the individual.

Protective activities – When an employee is leaving our organization, we need to recognize the potential damage the leaving employee can cause to the organization as he/she still has access to the sensitive information of the organization. The risk is higher directly in proportion with the rank of the employee in the organization. In some cases, like managers in the departments of finance and systems and network administrators, who have access to sensitive information of the organization, the risk is substantial. The higher the rank, the higher the risk. Therefore, we need to take these protective actions, especially in the case of senior managers and those that have access to sensitive information:

1. Isolate the person by removing access to sensitive information. That individual should be prevented from making decisions. But since the individual would also have ongoing activities, we need to immediately designate another person temporarily, at least, to take charge but allow the leaving employee to act in an advisory role to the designated person.
2. We should not assign any new activities to the leaving employee and allow that person to complete the pending activities already assigned to him/her. If it becomes essential to assign new activities, then we need to take care not to assign important or sensitive activities to that person.
3. As far as possible, we should not create a situation in which the leaving individual needs to represent the organization with important personnel like the customers or the government officials or someone who is important from the standpoint of the organization.
4. Restrict the access to sensitive corporate assets like the processes, customer lists, client orders and other information, vendor lists, employee lists and any such other important information. This prevents the corporate information being leaked to outsiders.
5. We also need to restrict access to information about the positive events or negative information. We need to be especially careful about the information about any litigation that our organization may be involved in.
6. If the person leaving has access to the information from our research department, we should ensure that the employee would not be able to leak information about our research projects.
7. We should make all arrangements expeditiously to relieve the person as soon as possible. We may even go to the extent of paying compensation to the leaving employee, if necessary, for releasing the employee before the notice period expires.
8. We need to take any other situation-specific action that is necessary to protect our organization from the possible damage that can be caused by the leaving employee.

Activities that foster goodwill – Past employees are ambassadors of our organization. They can be ambassadors of goodwill or ill-will for our organization. There are somethings that would cause the person to be either a goodwill ambassador or an ill-will ambassador but that individual would be an ambassador, make no mistake about it! What took place before the person ended up resigning, we would never truthfully know and would not be able to reverse the negative impression but we can treat that individual in a nice manner and try to impress that person positively so that the leaving employee would be a goodwill ambassador for our organization. At this stage, the important things that the leaving person would expect from the organization are a clean relieving letter that can be used as a reference letter in a future attempt to get a suitable employment; clearance of dues due to that individual including terminal benefits and residual salary payable to that person; and the smooth passage through the relieving process. If we can ensure that these three things happen, then we have motivated that person to be a goodwill ambassador. We need to focus our efforts to direct the ill-will the employee carries is directed toward individuals in the organization but not toward the entire organization. The leaving individual could have had differences of opinion with some people but not with the organization. We need to bring that point to the notice of the leaving individual so that the organization gains a friend once he/she steps out of our portals.

Final actions – We need to perform certain activities before we can issue a relieving letter to the leaving employee and send him/her from our organization portals giving a warm send off and bid farewell. Here are the activities.

Clean resignation – We need to repossess the organizational assets and submit them to the concerned agencies. This action needs to be performed by the leaving individual. The leaving employee needs to obtain a no-dues certificate from each of the departments with which the individual interacted with. We need to update the personal records with all the information and if the employee is slated for a possible future employment, we need to store that information for future use. We should also initiate actions for payment of the dues including the terminal benefits, residual salary and any other payments that may fall due to the leaving employee. Finally, we should set the date of release, in consultation with the head of the concerned department from the pending work perspective, and the finance department from the perspective of making the final payment. We need to prepare the relieving letter and other letters like the experience letter, conduct certificate/reference letter and any other applicable documents that we need to provide to the leaving individual conforming to our organizational processes.

For dirty resignation – In dirty resignations, the individual may or may not perform all the activities to be performed by him/her. We have to designate some individuals in consultation with the head of the concerned department and then through that individual, perform the activities to be performed by the leaving individual. The assets, to the extent available, need to be collected and deposited with the concerned agencies. Now, instead of no-dues certificate, we need to obtain the dues pending from the individual from all the concerned agencies and consolidate them. Then we would not be able to provide a clean relieving letter but if the individual demands a relieving letter, we need to provide the relieving letter but recording the dues and the reasons for relieving the person from the organization. If we do not record the dues that need to be collected from the individual, we would not be able to collect them later on. Irrespective of whether we provide the relieving letter, we need to delete the employee from the rolls of our organization.

For fleeing – The activities for the fleeing employee is similar to that of the employee who resorted to dirty resignation. The difference is that we would not provide a relieving letter and in addition we need to initiate necessary actions to trace that employee to recover our dues. Second, we have a responsibility to the employees working in our organization. Supposing the employee meets with an accident returning from work and meets death, some courts may hold us responsible to trace and pay all the terminal benefits and insurance benefits if any. Therefore, we need to lodge a complaint of missing-person with the local police. If the police locate the person, we may simply complete the relieving activities and if the person is not located, then we may treat the case as a missing person. When a person goes missing, the statues specify a certain waiting period before declaring the person as dead. We need to wait out that period, before closing the file. If the next of the kin approach us for payment of dues, we need to hand over the case to our legal cell to handle the matter and settle it.

For retirement – The activities of relieving a retiring person are similar to that of an employee going out with a clean resignation. Then we go a little further. We celebrate that event. We organize a party with his/her colleagues with after-dinner speeches that extoll the individual's contribution to the organization, offer a bouquet and finally drop the individual at his home. Some organizations provide a special identity card to allow that person to visit his/her workplace to satisfy his/her nostalgia. We would also arrange for the payment of the terminal benefits to be paid on or before the last working day of the person and if there is pension scheme in the organization, we would initiate all the necessary steps so the individual begins to receive the pension check as soon as possible. A retired person certainly is an ambassador of goodwill for our organization. The fact that the individual stayed long enough to retire with us attests to that fact. So, we should make every effort to ensure that the passage is as smooth as possible.

For death – Somebody dying while in service is a tragic event. That person is still young and therefore, his/her family is left orphaned and their kids, if any, are still not grown up. So, we need to handle the tragic event of death in a very sympathetic manner. It would be great if we can release some money to meet the immediate funeral expenses. That will go a long way in sending a positive message to the rest of the employees that the organization is caring and is interested in the welfare of the employees. The rest of the release activities are similar to that of the fleeing employees because the employee is not present to fulfill those activities. But the difference is that all those activities would be handled with a positive outlook and with sympathy.

For disability – We do not discharge an employee for minor disabilities. The disability must be such that, the employee would not be able to perform any activity. Whenever any employee meets with an accident or illness that would impair the capacity of the employee to work in his/her present job, we generally try to see if the employee can work in any other job even with the disability. Only when we are not able to fit that employee in any position in our organization, we plan to discharge that employee from our organization. In these cases, we need to treat the case with extreme sympathy and allow maximum benefits to that employee. If the accident or illness that caused the disability was while on duty, we may need to pay special compensation adhering to the applicable statute. Since the employee would not be able to fulfill all the formalities of getting released, we need to perform those activities but with sympathy and a positive perspective.

For termination/dismissal – For termination/dismissal, the discharge is immediate as we do not like to have the person on our premises because that individual can cause damage to our organization. The person would be escorted out or our premises by our security personnel. Of course, that individual would be allowed to collect his/her personal effects but under the supervision of the security personnel. We perform all the activities of releasing the employee and if there are any dues from the terminated person to our organization, we will be initiating separate action in consultation with our legal department. Termination and dismissal are severe actions and we do not easily resort to them. First, it will send a negative message to the rest of our employees that the organization is not sensitive to human frailties and can cause some immediate resignations or in the near future. We ought to proceed through all the steps of the set procedure like, suspension from duty, charge sheet, and enquiry, allow a reasonable opportunity for the individual to explain and redeem himself/herself. Once a decision is made, we need to serve the individual with the notice of termination or dismissal and escort the individual out of our organization immediately. That individual would harbor ill-will toward our organization and would be our ill-will ambassador. We just have to accept that and move forward.

12.6 Release Activities

While different organizations follow different practices in releasing resigning employees, here are some activities that are usually followed across many organizations:

1. Resignation letter – This is applicable in the case employees leaving our organization on their own volition. For cases of death, the receipt of intimation becomes the point of reference for the release activities. In the case of termination/dismissal, the letter of dismissal/termination acts as the initiator of the release activities. All of these are basically written information, from the standpoint of the HR department that the employee needs to be discharged from the organization. All these are legal documents and need to be carefully drafted in consultation with our legal department or our counsel. Resignation is of course, originates from the leaving employee. All we need to do is process it and its legal ramifications are the concerns of the resigning employee. Resignation letter needs to be approved and forwarded to the HR department by the direct supervisor and the departmental head before it can be processed.

2. Exit interview – This is applicable only in the case of employees leaving the organization through the resignation route. Exit interview is usually conducted by the HR department by a trained person to elicit as much information from the individual as possible about the reasons for leaving and the organizational processes as well as any suggestions for improvement. This can be conducted either immediately after the resignation letter is received or just before releasing the employee. Each organization would have a specified format to record the minutes of the exit interview and a checklist to ensure that all aspects are covered.

3. Repossessing organizational assets – Each employee would hold some of the organizational assets which include information in documents or

manuals, hardware and software tools needed for performing the job, workstations like computers, laptops, cell phones and so on. Before an employee can be allowed to leave the organizations, we need to repossess all these assets and deposit them with the designated agencies. In the case of employees resigning from the services of the organization, the employee will do this and deposit them with the designated agencies and take a no dues certificate from them. Only the identity card is the asset that we possess at the last minute at the gate of our organization through the security personnel.

4. No dues actions – We provide a number of facilities to our employees. We give loans; we provide a thrift store to sell various items at discounts and on credit to be payable from the salary deductions. We might have paid some advance for performing some official work. All these have to be recovered from the leaving employee. So, we ask the resigning employee, to obtain a no-dues certificate from all those organizational agencies to which the employee may owe money. This will ensure that the leaving employee would not have any dues to any of our organizational entities that need to be collected after his/her departure. For other categories of leaving, we need to collect these certificates that may include dues of the employee to the organization.

5. Payment of dues – The organization needs to pay some terminal benefits and other payments to the employee. Most organizations arrange to pay these monies on the last working day of the employee. We need to pay the salary to the employee for working from the last pay day till the day of leaving. We may need to pay any long-term saving schemes our organization may be operating and the leaving employee may be a contributing member of such schemes. We may be operating some cooperative credit societies and the share capital paid by the employee needs to be returned. And in this manner, there may be other payments due to the employee. We need to collate all such dues to the employee and make arrangements to pay them on the last working day of the leaving employee. One note of caution: we may deduct any of the dues the employee owes to our organizations from the payments due to the employee. We should deduct what is owed by the employee from the payments due to the employee before making payments.

6. Re-employment options – When employees resign and leave our organization can be re-employed at a future date if there is a suitable position available and the employee is willing. So, when the employee is leaving our organization, we need to see if the employee is willing to come back

later if the conditions are mutually acceptable and make a note of it in the exit interview. Sometimes, the employee may be leaving us in conditions that are not to our liking and we may not like the employee to come back. Still, it would be advantageous to have this information because the future is uncertain and in organizations, we need to be prepared for the unpredictable and uncertain future. Who knows, we may need that specific employee for one of our future projects!

7. Relieving certificate – Relieving certificate is the final step in relieving a person from the organization. It simply records that the named person worked in our organization and other facts like the leaving person's designation at the time of leaving, his/her last drawn salary and if there are any pending actions against the person and other situation specific topics. If there are any monies due from the employee, we need to list them in this certificate and if we do not, we cannot recover them later. If there are any allegations against the leaving employee and if we contemplate any action later on, we should also include it here. If we do not include such aspects in this certificate, we cannot do anything about them later on. Relieving certificate is a "certificate" and would be used as such by the person receiving it. Other organizations that may employ that person would place trust on this certificate while hiring him/her. So, we should be careful while preparing and issuing the relieving certificate.

Having discussed the relieving activities, let us now discuss the issue of attrition planning

12.7 Attrition Planning

Can we really plan attrition? How do we ever know who is going leave us or if any of our employees is going to meet the unfortunate event of death or disability? We can never know. But every industry has attrition and we can obtain the average yearly attrition percentage for our industry from our industry association or from trade reports and business analysis reports published in trade and business journals periodically. Of course, this is specialized information and we may need to pay for it. We also know the general attrition percentages speaking with our co-professionals when we meet them in conferences and seminars. The attrition percentages differ by the class of industry, size of the organization, the stage of the industry, the reputation of the organization and so on. For example, the attrition is lowest in the government compared to private sector. The attrition is lower in manufacturing sector than in other sectors. The attrition is higher in the services sector than others. Even in services sector, the attrition is higher in customer-facing organizations than in backend processing organization. Again, there will be higher attrition in the professions much in demand than those in lower demand.

So, in this manner, we can make some predictions of possible attrition in our organization. For example, in software development industry, there would be higher attrition in the people with the expertise in the latest development platform. Sometimes back, COBOL programmers were much in demand; later on, it was the C language, then there were platforms like the SAP, then it was Java which had high demand and commanded highest pay rates. There will be something else in demand, in the coming days!

The first step in attrition planning is to enumerate the individuals that are going to retire in the coming financial year. This information is in our records and we can get it accurately.

The second step is to derive a total number of possible attritions, by applying the industry average attrition percentage to our headcount. We know this number hardly ever comes to reality but the very nature of averages is such. But it gives us a possible number of people we may like to lose through resignations in the coming year. From this figure, based on the market demand, we can estimate the number of employees by skill set that are likely to leave us in the coming year. The direct supervisors would be able to predict with some accuracy the number of people that are likely to leave them with the help of their close interaction with their subordinates.

Then prepare an attrition plan using the format given in Exhibit 12.1.

Now, we include this in our organizational over all HR plan of the organization. We can also ask the departmental heads to prepare their attrition plan and consolidate it. But they would only be able to give the numbers and perhaps details of those they suspect of being ready to resign and leave our

Exhibit 12.1 Attrition Plan

	Our Organization	Industry Average Percentage	Possible Attrition	Replacement Needed
Organization total	1000	15	150	125
Skill set A	250	30	75	75
Skill set B	120	10	12	10
Skill set C				
Skill set D				
Totals			237	210

organization. But that is also valuable information because we may be able to give some indirect input to those people and try to retain them or we may be able to put some understudies with those people to be mentored so that our organization does not suffer from their leaving us.

All in all, attrition is a reality in organizational life and we need to prepare our organization for this eventuality and ensure that our organization is impacted only to the minimal extent and the operations of our organization continue to proceed smoothly without any interruption.

13

HR Department

13.1 Introduction

When any organization is incorporated, the HR department is one among the essential departments that have to be set up first. Whenever an unfortunate event like closing down a corporate organization happens, HR department is among the last to be shut down. That is the importance accorded to the HR department by the statute. HR department (HRD) is not a revenue earning center. Rather, it is a cost center. It is a support department. In management parlance which classifies organizational departments as "line" and "staff" departments, closets HRD with other staff departments. Of course, all departments are equally important for the survival and growth of the organization; HRD is close to the hearts of the employees of the organization as HRD is the representative of the employees at the management table in all matters. It is rather their advocate. HRD not only ensures that human resources are available for all organizational activities at the right time and cost in the required numbers, it also ensures that all employees receive fair compensation on par with their peers in the industry and their working conditions and treatment are on par with those of their peers in the industry. So, the employees love the HRD mostly and hate it when things are not to their liking.

13.2 HR Department Functions

First, let us consider the functions to be performed by the HRD before trying to understand the best way to organize these functions. The following are the functions to be performed by the HRD:

1. Recruitment – functions include,

    (a) Collating requirements – resource requirements are raised by various departments and the HRD collates all such requirements and consolidates them. In so consolidating, the HRD would combine

291

together similar requirements for the purpose of soliciting resumes including advertising, if necessary.

(b) Sourcing resumes – HRD would source resumes from all the available sources against all the requirements. HRD would maintain a database of all sources of resumes including consultants, internet groups, advertising avenues and so on. Sourcing is also done through the present employees. HRD receives resumes throughout the year through the organizational web site and also at the gate.

(c) Shortlisting candidates for interview – HRD will review the available resumes against the requirements and then make a list of eligible candidates. Then the resumes would be handed over to the originators of the resources requests to assess the technical eligibility of the candidates. The purpose of HRD scrutiny is to minimize the burden of technical people by reducing the number of resumes needing technical scrutiny.

(d) Arranging technical and aptitude interviews – Arranging interviews involves coordinating with the candidates and the technical interviewer for a mutually suitable date and arrange facilities for conducting the actual interview.

(e) Selection and offers – Once the candidate is selected, HRD will arrange all the necessary approvals and then prepare the offer letter and gets the candidate to accept the offer. If one candidate rejects our offer, HRD may negotiate with the candidate in consultation with the technical team and if the negotiation did not succeed, another candidate may be offered. HRD coordinates all this process.

(f) Induction of the employee – Induction of the employee includes ensuring fulfillment of all the joining formalities including filling and signing all the required documents, introducing the new employee to the concerned persons, conducting the induction training as needed and so on.

(g) Hand over to the technical team – This involves, introducing the employee to the supervisor under whom the new employee would be working and other colleagues in the department and hand over relevant documents to the supervisor.

(h) Interns – Interns is a facility provided by the organizations to the nearby educational institutions and this will enable the students to get a firsthand feel of real-life application of theory they learnt in the colleges, to practice besides giving them a few extra Dollars to meet their expenses. While interns carry out some work, it is usually not of the same quality as of the experienced employees. So, it is a sort of expense to the organizations as the value of the work completed by the interns may not meet the expenses of employing the interns. But it also allows us to closely observe a qualified person for much longer periods than in an interview and allows us an opportunity to earmark that person for future employment. Organizations select the interns and offer them employment on completion of their degree. It is advantageous because the intern becomes immediately productive on completing his/her course and joining our organization because of the previous experience the intern had in our organization during his/her internship. HRD needs to coordinate the entire program of internship. This includes coordination with the educational institutions, selecting the right candidates for internship, facilitating the work of the intern in our organization, coordinating with the technical teams for the details of the work carried out by the intern and then, finally relieving the intern from our organization with a certificate of internship. It will also include making offers and getting the selected intern to accept our future employment offer at today's terms. HRD would be the guardian of the intern, during his/her stay with the organization.

(i) Job fairs – job fairs in the recent times have become popular for recruitment of right candidates. Job fairs are the meeting places where a number of employers and those seeking employment come together. Generally, the organizations set up a booth and display the available vacant positions, provide information about the organization, working areas, general compensation and benefits and so on. In those booths, HRD persons would be present and answer any questions of the candidates. Candidates can submit resumes and if the employer representatives feel that it is a good resume, that candidate may even be interviewed preliminarily and be invited for an interview with the technical team. Job fairs are organized by educational institutions, industry associations, consultancy organizations and so on. Job fairs are also an opportunity to meet co-professionals from other organizations to exchange and collect information about various practices and processes, compensation patterns which can be put to future use. HRD needs to take ownership of the entire work involved in attending as well as conducting the job fairs as and when required.

2. Payroll administration – Administration of organizational payroll involves making various employee-related payments. While it is the finance department that actually makes all the payments, HRD sets the policies and various rules to be adhered to in making all such payments. Here are a typical set of tasks administered by the HRD.

   (a) Attendance, leaves, holidays – Attendance is monitored by the HRD. Any absence by the employee is monitored and ensures that the information is fed to the finance department for necessary action. If the absence was authorized, then the finance department would be advised not effect any deduction. The absence can be due to official work assigned to the employee outside our facility which may include assignments such as onsite work, official tour, attendance to training programs, workshops, conferences, seminars, client meetings and so on. Similarly, all leaves of absence taken by an employee are subject to the final approval of HRD. The leave of absence can be within what is allowed for all employees as organizational policy or some sort of special leave like the compensatory off for working extra hours earlier. Or it can be LOA without salary for special circumstances. HRD sets the policies and then administers them. Every organization allows some holidays every year based on national holidays, festive holidays and such other holidays. These holidays are declared at the beginning of the year in consultation with the organizational management and employee unions, if existing.

   (b) Salary payment – While the actual payment of salary is made by the finance department, it is administered by the HRD. These payments include the regular salary based on the hourly-rate agreed with the employee, overtime payment if any, incentive payments, if any and then implement the deductions like absence, insurance payments, any checkoff from the salary and other deductions authorized by the employee. Salary for every employee is set by the HRD initially and any changes like the annual increments are also decided by the HRD. HRD informs the finance department of such changes for implementation. HRD ensures that the salaries are paid on time on the set dates.

   (c) Payment of benefits like bonus, incentives, advances, reimbursements – There would be many one-time payments like the bonus, incentives, awards, project-end bonuses, reimbursements for expenses incurred for official purposes that become payable to employees from time-to-time. All these would be decided by the HRD and informed to the finance department for implementation.

(d) Recoveries – There are a variety of recoveries to be deducted from the employee salaries. These include court-ordered payments such as child support, alimony, loan installments; then employee authorized payments like the union membership fee, club membership fee, professional association membership fee, shopping from shops authorized by the organization, transportation charges where the employees are fetched on an organizational transport facility and so on. All these are administered by the HRD. Sometimes the deductions exceed the salary payable to the employee! In such cases, HRD will ensure that a minimum amount is paid to the employee adhering to the organizational policy and manage the loan payments in such a way that the employee interests are protected.

(e) Compensations for accidents – When an employee meets with an accident while carrying out official work, the organization becomes liable. Usually, the organization takes an insurance policy for such exigencies. HRD is the agency that coordinates collection of the money from the insurance company and makes payment to the employee. We would pay some money immediately to the employee to meet immediate expense and recover it from the claim money received from the insurance company later. HRD would make the claim, follow it up till receiving the payment, disbursing it to the employee after adjusting the advances, if any, paid to the employee.

(f) Retirement plans – Every professional organization provides for a retirement plan for its employees. Mostly, governments run such schemes and organizations take advantage of these schemes. In fact, infant organizations that were recently set up are not allowed to run their own retirement plans. Such companies have to use the governmental schemes. Mature organizations, however, are allowed to set up their own retirement plans but subject them to governmental oversight. HRD is vested with the management and coordination of all retirement plans including compliance to statutes in force.

3. Employee relations – This function was earlier referred to as establishment or industrial relations. Establishment referred to those actions which were focused on implementing the rules and regulations without

any bias in the organization. Then when unions came into existence, maintaining cordial relations with the employee unions to avoid and avert strikes establishment became important. The name was modified to industrial relations or simply IR. With the pressure from employee unions easing, and the organizations realized that they have to ensure that employees are kept satisfied to ensure productivity and quality. So, the function metamorphosed to employee relations indicating a quantum shift in the organizational outlook in the matter of employee satisfaction. The following functions are performed under this umbrella now.

(a) Occupational health – Organizations have come to agree that employees can suffer health problems merely because they work in the organizations. So, they took it upon themselves to monitor the employee health and ensure that they are not suffering from health problems due to the work they perform in our organization. The health problems may include mental health too. Organizations undertake a periodic health assessment in areas where there is a risk to health. They also provide a contributory health insurance to all employees in which major or entire portion of the health insurance premium is paid by the organization. This is one of the important functions carried out under employee relations.

(b) Travel – Whenever employees needed to travel on official work, the organization needs to make all arrangements for travel including booking tickets, hotels and so on. When organization undertakes this work, large discounts are given to corporate users in antic-ipation of bulk business. To take advantage of such discounts, organizations do undertake this activity.

(c) Holidays – Yearly holidays need to be decided before the new year begins. This has to be done in consultation with employee representatives, the management and the governmental agencies. For giving delight to the employees, sometimes, long weekends are provided by giving an extra holiday on working day that falls in between two holidays.

(d) Working hours – We do not pay much attention to this matter in these days because, it is more or less a norm to work between 8 AM and 5 PM every day in single shift organizations. But there are organizations that work in shifts. They may work 2 or 3 shifts a day. In those cases, it becomes necessary to decide the shift timings in such a way that the employees feel comfortable working in those timings as well as keeping their health in their normal conditions.

(e) Home working facilities – In the present times, quite a few companies are resorting to wok-from-home mode of performing organizational work. The organizations have come to the realization that it is not necessary for employees to gather at a common place to do the work. While this began with software development companies and IT (information technology) departments, it has now spread to many other departments as well. Any work that does not need costly equipment to perform work, organizations are trying to see if the jobs can be performed from the employee homes. After all, when work can be outsourced, can it not be home-sourced? It is allowing significant and visible cost reduction in the terms of reduced rent, energy costs, communication costs, facility maintenance costs and so on. Now, it is the employee relations department which facilitates and coordinates the work-from-home of the employee homes, especially taking care of the human side of it. The technical departments arrange technical facilities so that the employees can perform the work from home.

(f) Grievance resolution – In our humble opinion, there is no organization in which no employee has any grievances. Some excellent employee-focused organizations have minimal grievances and some poorly organized organizations have more grievances. But all organizations face grievances from their employees. These grievances need to be resolved to keep the concerned employee satisfied. Organizations have grievance resolution procedures which are implemented by the employee relations function of the HRD. HRD is the agency that receives the employee grievances and tracks them through to resolution.

(g) Retention and attrition – While the employees work in all the departments of the organization, HRD takes ownership of all human resources of the organization and champions their cause. HRD puts in place all necessary processes to ensure long tenures for all the employees in the organization and in coordination with all departmental heads ensures retention of the employees. When the unfortunate event of the attrition of the employee happens, HRD takes all actions to ensure that the organization derives as much benefit from the event for the organization by extracting as much useful information from the departing employee as possible.

And all the actions including the final release of the employee from the organization are handled by the HRD.

4. Skill retention and development – Skill retention and development include all activities that would ensure that our employees are on the cutting edge of their own field. HRD takes ownership of this activity and champions the cause within the organization. Here are the coordination activities performed by the HRD.

   (a) Collation of requirements – the skill retention and development requirements are collected from all the organizational entities, consolidated and the identified candidates are sponsored for various programs.

   (b) Planning of training – After consolidating the requirements from all the organizational entities, HRD would prepare an organizational training plan enumerating the training programs required to be conducted. Then, HRD would identify the programs that can be conducted in-house and the programs to be conducted by outside programs or public programs. Then based on this information, prepare an organizational training calendar and obtain approval for the same from the organizational management. Once it is approved, HRD would implement the plan and achieve the objectives set for the same.

   (c) Sponsoring to public programs, conferences and seminars – Some of the futuristic skill development programs need to be met by sponsoring our employees to various scientific organizations and professional associations. HRD would collate these requirements from various organizational entities and prepare a budget and obtain financial sanction for this activity. Then HRD would look for suitable opportunities to sponsor the selected candidates for these conferences/seminars. When such opportunities are sighted, HRD would take necessary actions to sponsor the candidates selected in consultation with the departmental heads and management.

   (d) Conducting in-company training programs – Whenever any in-company training program is to be conducted HRD would take ownership of such programs and coordinates the entire activity. It would arrange the classroom, training facilities, arrange the faculty and arrange the equipment needed for hands-on sessions. When the training program is underway, HRD representative would handhold the program till it is successfully concluded. The follow up actions like taking feedback from the faculty and the participants and any other actions to assess the impact of the training program would be taken by the HRD.

(e) Repository – HRD usually maintains a repository of the course material, training material, audio and video recordings of the past training programs for future use. This repository can be used by our employees for self-study and acquiring the needed skill or bridge the skill gaps.

5. Statute compliance and reporting – A professional organization has a host of regulations that it has to comply with and report such compliance to statutory authorities. Some of these relate to the product, the facility, pollution control, noise, hazards, safety, security, export/import, non-discrimination, social responsibility and so on. Different agencies are responsible for different statutes. HRD is responsible for compliance to all statutes relating to employees, their safety, safety, non-discrimination, equality of opportunity, working environment and so on.

(a) Safety and education of safety procedures – Workplace safety is organizational responsibility. HRD takes ownership of employee safety. As part of employee safety, HRD needs to ensure that all the employees are appropriately educated in the safety procedures implemented in the organization and keep all necessary records for auditing by the statutory authorities if the need arises. If any unfortunate accident takes place within the organization or outside the organization while the employee is on duty, HRD takes the responsibility to ensure that all stipulated actions are taken and the suffering employee is treated and compensated as specified. HRD also has the onus to report all such incidents to the statutory authorities. HRD analyzes all such incidents and implements necessary improvements in the process, equipment, tools and education to prevent recurrence of such events in the future.

(b) Security – Security has two sides. The first side is the security of the organizational assets and the security of the employees. While the employees are on duty while inside the facility or outside of it, the security of the employees is the responsibility of the organization and the HRD takes the onus for it. In many organizations, some of the work is risky. For example, a driver of a bus fetching the employees faces the risk of accidents. A machine operator faces the risk of injuries. Employees sitting in the office may face the risk of the roof caving in and falling on their heads. There can be an earth quake. Some of the risks are avoidable by appropriate equipment, necessary maintenance of the equipment and tools, education to the employees about the risks and how to mitigate them and so on. But security incidents do happen occasionally and when such events take place, HRD needs to swing into action and take all necessary actions to ensure that the risk to the employees is minimized. Of course, HRD needs to equip itself with necessary equipment, insurance policies, people and processes necessary to handle such situations effectively.

(c) Emergency preparedness – Emergencies do take place in the organizations, occasionally. Events like earth quake, terror attacks, fires and so on bring on emergency situations. HRD fronts the organization in such situations with the help of other departments especially in respect of the employees. HRD periodically conducts audit for emergency preparedness to ensure that everything necessary is in place. When such incident happens, HRD accounts for the safety of all employees and takes appropriate action for the affected employees.

(d) Insurance – HRD is the coordinating agency for all insurance policies taken by the organization. An organization takes insurance policies for medical, property, accidents, general liability, fire, floods, earth quakes, terror attacks, and so on to ensure that none these incidents can cripple the organization. While the finance department pays the premium for all these policies, HRD manages these policies and ensures that all are current as well as make claims whenever necessary and collects the money as claimed.

(e) Accidents/security/safety incident response – Whenever any unfortunate incident takes place in our organization in respect of accident/security/safety and so on, HRD spearheads the response, coordinating and directing the incident response teams. It captures the details of the incident for future analysis and assists the investigation. HRD coordinates and cooperates with the investigating and statutory agencies and ensures that the culprits, if any, are caught and brought to the book.

13.3 Factors Influencing the HRD Organization

Now, having discussed the functions performed by the HRD, let us now discuss the organization of the HRD which depends on several factors based on the organization, such as,

1. Size of the organization
2. Single building or multi-building
3. Geographical spread within the country
4. Multi-national organizations

Now let us discuss the HRD organization for these organizations.

1. **Size of the organization** – As can be easily perceived, larger the number of people, we need a more elaborate organization. A small organization would have multiple functions combined into one role and one person handles all such combined functions. Having a separate individual handling each function would not be financially viable for a small-sized organization. A medium-sized organization would be able to afford one person per function and some of the related functions could be merged into one to save on the expenditure. A large organization would be able to allocate more funds to the HR function and therefore would be able to employ fully-qualified experts to handle each of the above functions. Thus, the size which determine the financial capability to spare adequate funds to HR function would have a bearing on the organization of the HR department.

2. **Single building or multi-building** – A single-location organization may be housed inside a single building or it could be spread over multiple buildings. If it is a single building organization, then the department can be centralized and located in one place. If it has multiple buildings, then it needs to have some amount of decentralization so that each building would have a representative of the HR department to service the employees in that building and interface with the central HR department where necessary. Alternatively, each building could be treated as a small independent profit center, in which case, each building would have a self-sufficient HR department and a central HR department would set policies to bring in uniformity in HR practices across the organization. There would be an individual handling the HR function independently in that building but within the framework set in place by the central HR department. There could be alternative practices in vogue in disparate organizations. Thus, the number of buildings in the organization would influence the organization of the HR department.

3. **Geographical spread within the country** – An organization may have several facilities geographically distributed but within the country itself. In which case, each facility needs to have a separate HR department but based on the size of the organization at the specific location. Some of these facilities can be large, some medium and some can be small sized. So, at each location depending on the size of the facility the HR department needs to be organized. In addition to these HR departments at the individual locations, there ought to be a central HR department which puts the overall framework and HR policies in place to be implemented across the organization to achieve uniformity of HR practices followed in all the organizational units.

4. **Multi-national organizations** – Multinational organizations are very special in the sense that each organizational unit needs to have a set of HR policies and practices depending on the country in which it is located in. The salaries and wages have to maintain parity with those that are being paid in similar industries and sectors in that country. The holidays need to be based on the festive days of that specific country. Similarly, every aspect of HR practice needs to maintain parity with those organizations that are similar in nature with our organization. In such a case, what kind of uniformity can be maintained across the organization? Obviously, we cannot but we need to ensure those principles that every organization cherishes must be maintained across every organizational unit irrespective of which country it is located in. If we have multiple units in different countries, we need to maintain uniformity of HR practices and policies across all units within any given country with added local flavor.

13.4 HR Department Organization

Now having considered every aspect of organizing the HR department, let us now discuss the organization itself. Let us fix positions for a full-fledged HR department so that it can be scaled down to suit the smaller organizations.

HR department would occupy, usually, the same position as other main departments such as the marketing department, delivery department, finance department, research and development department and so on. HR department is not subordinate to any other department and generally reports to the CEO of the organization or unit as the case may be. Figure 13.1 depicts an organization chart showing the position of HR department in an organization.

Figure 13.1 Position of HR in the organizational hierarchy.

First and foremost, we need an individual to take complete responsibility for the organization's entire HR function. Most organizations place this person on par with other departmental heads including marketing, finance, technical, quality assurance and others. This individual needs to have a post-graduate qualification at a minimum, and put in a number of years of experience in the HR department of a professional organization, including ours. In our humble opinion, a qualified individual should have put in a minimum of ten years of experience before that individual can be considered for heading this vital HR function. Some organizations compromise on the experience if the individual obtained the qualification from a premiere educational institution or university. Again, in our humble opinion, any amount of education from any premiere educational institution, can never supplant practical experience. The position of the head of a vital and strategic department like HR can either elevate the organization toward excellence or drive it toward extinction. Dirtying one's hands in the field gives the much-required maturity to the individual and prevents that person from taking hasty decisions detrimental to the organization's interests.

This individual needs to have experience in all fields of HR functions, especially, in recruiting, employee relations and salary administration. Even in these functions, that individual might not have performed all the sub-functions by himself or herself, but he/she should have worked in those functions. That way, even if the individual did not perform all the functions, but would have had an opportunity to observe all the functions being performed by others to learn. At a minimum, the individual ought to have put in a minimum of two years in each of those functions.

Other desirable personality traits in the person would include, but not limited to, high level of integrity, empathy, negotiation skills, fairness, goal oriented and problem resolving skills in addition to those skills that every manager needs to possess of which we wish to emphasize communication and interpersonal skills.

Some organizations prefer that the head of an HR department needs to have a degree in law but in our humble opinion, it is superfluous as we would any way have a legal department or an organizational attorney who would advise as necessary on legal matters. It would suffice if the individual has a postgraduate qualification in the management with a specialization in HR management.

Next position we need to consider is the manager for recruitment. The position of recruitment is critical to the success of the organization as its main function is to provide the inputs, that too vital inputs of human resources to the organization. While this person can have similar qualifications to that of the head of HR department, we can give some relaxation in the matter of qualifications and experience. This person needs to have worked in the recruitment area either at a similar organization or a staffing company and must have put in a minimum of five years of experience if drawn from a premiere educational institution or more years if from a normal educational institution. Among the personality traits this individual needs to possess, we emphasize the trait of integrity as it is very important. It is easy for this position to bring in second and third-rate human resources under various pressures the environment subjects this person to. This person needs to have all the personality traits that the head of HR needs to have.

Next position is that of manager of payroll administration. Payroll admin-istration is a very important function of the organization as errors in the payments can very quickly cause dissatisfaction to the employees. In fact, a single error in the pay can cause the employee to lose focus until that error is rectified. Therefore, this individual ought to be the one who is always meticulous and diligent. Since the work is repetitive, we need to select this individual such that the person can carry out fairly routine but essential and critical work without feeling the monotony. The important personality trait needed in this individual is that of diligence. The academic qualification needed of this person is a degree in management.

Next position is that of manager of employee relations department. Manager holding charge of employee relations department is critical in ensuring that the employees are satisfied with the organizational policies and procedures and the organization runs like a well-oiled machine smoothly. He needs to have similar qualifications as that of the recruitment manager and experience in an employee relations department for about five years at a minimum if the individual was qualified from a premiere educational institution and more years of experience if that individual was qualified from a non-premiere educational institution. In terms of personality traits, high integrity and interpersonal skill take the premium spot along with other traits needed for a manager in the organization. As this person needs to keep a tab on the pulse of the employees, he/she needs to be an outgoing individual who loves to interact with other employees.

Next position we need to consider is that of the statute compliance manager. It is better that this person is qualified in labor law or something similar or a basic degree in law. It is counter productive to recruit a law graduate from a premiere educational institution because such persons would quickly get bored with the limited scope for application of the legal skills learnt at the college and would certainly quit. We can even do away with the requirement of law degree as any management degree with specialization in HRM, would include the rudiments of labor law. This person must be meticulous with an eye on the detail to oversee all aspects of organizational functioning from the standpoint of applicable statutes and bring out gaps to the notice of the management for correction and compliance.

Next position is that of manager of skill retention and development. We need an academic person here as the work is basically teaching the employees how to do their work better and it is like teaching the fish how to swim better. The employees know their role and responsibilities. What this department does is to prepare them for the future and train them to do their work using better tools, better techniques and better methods as well as to expose them to new developments in their field. This individual needs to be an academic with an interest in our field of endeavor and research. We need to recruit a person that has curiosity as his/her dominant personality trait so that he/she continuously scans the environment to notice new developments in our focus area and bring it to the notice of the concerned management people.

Figure 13.2 depicts an organizational chart of an HR department for a large-sized multinational organization. We can scale it down depending on various factors, discussed earlier in this chapter, that affect the organization of the HR department.

Having considered the functions of the HR department and the managerial positions needed, let us now, discuss about the quantum of staff needed to fill the positions. In smaller organizations, it is common to combine the positions of employee relations with that of the head of the department; and recruitment position and skill retention and development into one position usually payroll administration in small organizations is offloaded to finance department. We can find various permutations of this type in different organizations. Let us now look at the organizational chart from the standpoint of a large, single-location organization. For smaller organization, it needs scaling down. For multinational organizations, it needs replication across their offshore organizational units.

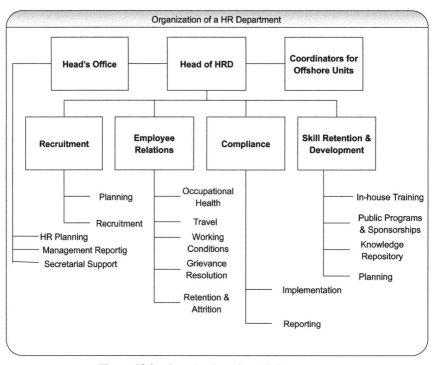

Figure 13.2   Organization of an HR Department.

13.5 Staffing the HR Department

Now the question to be answered is 'how many people are required to run an HR department effectively?' It is very difficult to answer because, the workload placed on the HR department varies from organization to organization; the tools used in the department to carry out the work differ from organization to organization; there are a variety of methods adopted in the organizations; and the workflow itself varies from organization to organization.

Now, there are a variety of computer-based software tools and gadgets. Now the workload of time-keeping has come down to a minimum with clocks connected to the computer, flexible working hours and home-working. Workflow of absence management is taken care of by software. The typing and preparation of various letters including offer letters, appointment letters, promotion letters and so on are produced automatically using software tools. Now the HR executives are able to perform much more work, with better quality and shorter turn-around times than before. Computers and software tools obsoleted the positions of secretaries and assistants. Decision support systems eliminated the positions of specialists like the statisticians.

However, in medium to large organizations, we need at least one person for each of the functions. Then depending on the workload, we need to assess the requirement of people to staff the HR department.

Recruitment is an involved work needing more people to handle the work. But if our policy is to recruit people from campuses, then we do it once a year. For once a year recruitment, it is not cost-effective to employ more people on the rolls. Instead, we can draw people temporarily from within the organization or have temporary employees hired from consultants or even outsource the people-intensive activities to those consultants specializing in the recruitment. But, if we are recruiting experienced people from the job market round the year, then we need more people on our rolls depending upon the number of recruitment cycles. The recruitment cycle can take anywhere from one month to three months typically but could be shorter or longer depending on the position. We can safely assume that one recruitment executive can handle a minimum of six cycles of recruitment a year. Taking this as a norm, we can work out the number of recruitment executives needed in our organization.

Employee relations workload is directly proportional to the number of employees on our rolls. But if we use a robust HR software package as well as a robust payroll software, then the grievances and other workload would be kept to a minimum. In such a scenario, we can have one executive to handle the workload generated by 500 to 1000 employees. But if we are using manual methods or a mix of methods, we need to work out a different norm for our organization Using such a norm, we can work out the total number of people needed to handle the employee relations workload.

In the skill retention and development wing, the workload depends on the number of training programs we conduct and the number of sponsorships we send to external training programs and seminars. Each internal training program involves planning, selecting the participants, arranging the faculty and facilities and coordinating the conduct of the program. Generally, a training executive would be able to conduct 2 to 3 internal training programs concurrently. So, based on the number of internal training programs we plan in a year, we can work out the number of executives needed for conducting the internal training programs. External training programs involve less workload than internal training programs. It involves collecting the information about the public training programs being conducted in our area of specialization, circulate such information to relevant executives, receive nominations, obtain financial sanction, sponsor the executives fulfilling the requirements, make arrangements for the executives to attend the program and finally obtaining the feedback from them. The total amount of workload for each sponsorship, which may include multiple executives being sponsored to a single program, could take up to one or two days. Using such norm, we can work out the requirement of executives for handling the sponsorship to external programs.

In the handling of the knowledge repository, the role of this department is limited to coordinating the effort. So, we need at least one person to coordinate the entire effort and then we need one or two people to glean the publications and mark up the relevant portions and get those pieces vetted by the concerned technical people. If approved for inclusion by the technical people, arranging the entry of the selected pieces of information into our knowledge repository needs to be accomplished by the data entry operators either in our department or a central CIO (Chief Information Office). Therefore, we need at least two people, one to coordinate the information gathering and inclusion and one person to manage the repository as well as for dissemination of information. We would need one person to coordinate the effort for preparing and circulating the repository newsletter. If we need to prepare the knowledge repository newsletter within our department, then we need additional staff to prepare the newsletter. Normally, all newsletters for intra-company circulation are handled by corporate communications department in which case, the work of preparing and circulating would be handled by them. In many cases, this portion about knowledge repository would only be a section in the overall corporate communication newsletter to the employees.

For compliance section, we need at least one person to coordinate the effort. In fact, the compliance is to be achieved by the concerned technical departments. All that the HR department is required to do is to bring the information about new statutes or changes in the existing statues to their notice. Alert the departments when a statutory inspection is scheduled and coordinate with them to ensure that all necessary things are in place to get through the inspection without a hitch. Also, coordinate with the technical departments to obtain the information about the compliance matters periodically to prepare the statutorily required reports and submit them on time to the statutory authorities. One executive should normally suffice unless ours is an industry that has built in risks and dangers, such as a mining industry. In such cases, we would need more executives depending on the size and risks associated with our working conditions.

Head's office is needed to support the head of the HR department. A head of HR department needs to coordinate and take ownership of the entire HR function. Therefore, he needs various reports from all functionaries in his/her department as well as provide various reports to others in the organization. The head needs to perform various analyses about the functioning of the department as also how that affects the entire organization. Who should run this office besides the head? We need someone who is well versed in making various analyses to get the information desired by the head. This person ought to be adept in data research to come out with information needed by the head of HR department. That individual may need one or two assistants to help in the data collection, preparing the reports based on the data analyses carried out and information handling.

Coordinators for offshore units or geographically disparate units are needed when the organization has independent units at other towns including offshore units. In such cases, we need one executive for each offsite unit. If such units are very large in size, then we may need more than one person to handle that work.

13.6 Performance Measurement of HR Department

These are the days of computers and number crunching. Good impression of the boss is inadequate. We need metrics to prove our performance. How do we measure the performance of an HR department which basically renders service to the organizational units? If we recruit people, it is only when others have an approved requirement. We resolve grievances only if there are any! HR service is like that of electricity. As long as our appliances are working who would care to find out where the electrical substation is located or who takes care of our area? But when the power is cut, every one begins blaming them! If the organization functions smoothly, the credit goes to the technical units. But if there is a hitch, it is somehow due to the fault of the HR department!

But because HR department is a cost center affecting the cost, we may compute the metric of cost of HR per employee serviced. That is

HR cost metric = cost of HR department ÷ total number of employees

While we cannot prescribe a good metric, we can confidently say that it should be kept to a minimum. But can this metric be about 50% – obviously not! There is no consensus on this metric but in our humble opinion, it should be kept between 5% and 10% or the money spent on the HR department needs to be kept under 10% of the total cost of running the company.

Some people talk of ROI (Return on Investment) and it is rather fashionable to talk about ROI. ROI is computed by dividing the net benefit derived from the activity by the expenses and the ratio ought to be greater than 1. But HR is not a revenue earner and the benefits are intangible. It is advocated that we can assign some values to the intangible benefits derived by the HR department but in our humble opinion, it is rather a "guesstimate" than a real objective metric. In our humble opinion, that it is a subjective metric that can produce any desired value and is subject to human bias. HR cost metric discussed above amply suffices for measuring the effectiveness of the functioning of the HR department.

In recruitment, two metrics are popular, namely the cost per hire and turnaround time. Cost per hire is the total money expended in recruiting one new employee. It is calculated, thus,

Cost per hire = Total amount of money spent on recruitment in a year ÷ the number of employees recruited in the year

The cost needs to include the money paid to the candidates for travel, expenses paid to the external experts, cost of conducting tests, if any, and so on in addition to the salaries and expenses paid to the employees working dedicatedly in the recruitment work. We cannot specify an ideal metric for this measure as it depends on the organization. What we can say is that this needs to be maintained or reduced over a period of time.

Another important metric used in the recruitment is the turnaround time to recruit an employee. Turnaround time is the amount of time taken from the date the recruitment request is received to the date on which the selected candidate joins the organization. It is computed in the number of calendar days. It is computed thus,

Recruitment turnaround time = Total time taken for the recruitment requests completed in a year ÷ total number of employees recruited and joined in the year

Total time taken is the sum of the days taken for completing each recruitment request in the number of calendar days. The range is usually between 1 month to 3 months and the average time is around 45 days. Our metric needs to hover around this value.

For employee relations function, in our humble opinion, metrics like cost per grievance resolved etc. are not really applicable even though some organizations use it. Grievances arise not due to the HR department. So, if the grievances are not raised at all or too many grievances are raised is not the true measure of the performance of the HR department. But the turnaround time to resolve a grievance is relevant. We compute this metric using this formula:

Turnaround time for grievance resolution = Total time taken for resolving the grievances resolved in a year ÷ total number of grievances resolved in the year

The amount of time to resolve a grievance is the difference between date on which it is received and the date on which it is resolved in the number of calendar days. It is in calendar days. Normally most grievances are usually resolved in a week in most professional organizations. But we should not take that long for all grievances. We should keep our metric less than one calendar week.

For skill retention and development function, we usually use the metric of cost of training per trainee and the turnaround time to conduct a training program. We compute the turnaround time taken for conducting a training program using the following formula:

Turnaround time for conducting a training program = Total time taken for conducting the requested training programs in a year ÷ total number of training requests received in the year

The time is counted in the number of calendar days from the date the request is received to the date on which the training program is completed. The cost per trainee is computed using the formula:

Cost per trainee = Total amount of money spent on internal programs conducted in a year ÷ the number of employees trained in the year

The costs should include all expenses including the cost of the faculty which might have been drawn from internal sources or external sources, cost of course material and training material, the cost of the HR executives involved in coordinating the training programs, the fixed costs of the employees in the skill retention and development section and so on. What we do not include is the cost of the participants as well as the opportunity cost of those participants.

We do not usually measure the effectiveness of the knowledge repository and it is treated as an expense or cost of keeping our human resources on the state of the art in terms of developments in their field.

14

Roles and Responsibilities in Managing People at Work

14.1 Introduction

Managing people at work encompasses the entire organization and it is not limited to one department, HR. True, HR department owns the people at work but, all managers and supervisors as well as the top management are all involved in ensuring that people work optimally in the organization achieving the organizational goals together. HR certainly takes not only ownership of the organizational human resources but also takes the lead in formulating the processes and policies that govern how the human resources are managed in the organization. In this chapter, we will discuss the roles and responsibilities of all concerned agencies in the organization.

14.2 Stakeholders in People Management

Every one in the organization is a stakeholder in the aspect of managing people at work. Still, let us clearly enumerate them here so that we can discuss the role and responsibilities of each stakeholder:

1. Top management
2. HR department
3. Process definition and improvement group
4. All supervisors and managers
5. Employees themselves

Top management – Top management is responsible for everything in the organization including the setting up of the organization, organizing it, funding it, getting the work done, ensuring the quality of deliverables, marketing their products and the services offered by the organization and so on. One of those responsibilities is the ensuring that the people work effectively as well as ensuring that the employees are treated fairly. People come into the organization only when the top management approves the recruitment and provide funds to pay them. Since people were brought into the organization at their behest, the top management personnel are responsible for them. However, their role is in funding, initiating and approving various organizational policies that affect the human resources.

The main responsibility of the top management is to ensure that a framework is put in place in which the people management can flourish. We discussed about the framework required for effective people management in Chapter 2 of this book. Let us now, enumerate the responsibilities of the top management here:

1. To put in place a framework for people management in the organization that is appropriate for the organization.
2. Define right kind of HR policies for the organization.
3. To fund all the activities necessary to ensure that the activity of people management can proceed smoothly without any hurdles.

The main role of the management in people management is to ensure funding the activity on a continuous basis as well and to monitor the activity and to improve the framework using a process-driven approach. Let us enumerate the role of the top management in ensuring that the people management is carried out effectively:

1. Be the advocates of the importance of proper people management so that everyone in the organization takes the message and ensures that people are managed efficiently effectively and appropriately.
2. Monitor the people management regularly to show support to the activity – unless the top management periodically monitors the people management activity, the lower levels would not accord due importance to this activity which would result in poor people management in which the people would feel like they are treated as a pair of hands and nothing more. Regular monitoring of the activity by the top management prevents such a state.
3. Ensure that the defined processes are implemented across the organization without any laxity – While the process definition in organizations happens diligently, its implementation is often lax. The reason for such laxity is that the rigor of implementation sometimes comes in the way of expediting things especially when there is pressure of some sort. To the question which to choose in pressure, the

answer is difficult. We might have to say that we should avoid pressure-situations to begin with. But, the reality of organizational life makes it inevitable to have pressure-situations occasionally. However, a well-defined process would have waivers for such tight situations, and the line managers would take recourse to such waivers. It is for the top management to ensure that the waivers are kept at a minimum. Then every organization would have someone who abhors processes and champions unbridled freedom to the managers. Even though the arguments against processes and standards were settled in favor of processes and standards, there are still individuals that regurgitate those overused but settled arguments against the processes and standards. It is for the management to handle such personalities in the organization and ensure that everyone participates in the process implementation heartily or at a minimum, would not vigorously oppose or work against the process implementation.

4. Ensure that the defined processes are improved periodically adhering to a defined process for improvement – Resistance to change is inherent in people. There would be resistance to process implementation initially and then there would be resistance for improving those processes. There would be arguments like, "Why now?" or "I don't see any benefit emerging out of this improvement" or "It puts extra burden on our people" and so on. Again, the top management has the onus to convince such detractors and enable the necessary process improvement.

5. Accord approvals as necessary for processes, policies and other relevant artifacts as necessary – The word "approval connotes just a signature to common folk who do not have to approve anything. Some approvals, like approving the leave application of a subordinate may be like that but approval of process definitions and improvements is not so easy. For these things, there would be opposition, then there would be side-effects as there is no solution in the world without room for side effects and new problems besides the commitment of funds. People work enthusiastically to define processes tightly and then to improve them to make them even tighter. In their enthusiasm, people would not be able to see the side effects, the strength of the opposition or the requirement of funds. It is for the top management to moderate such over enthusiasm and bring pragmatism into the artifacts and then approve. The onus for such pragmatism and wisdom is squarely on the shoulders of the top management.

6. Participate in all people management initiatives to show the support the top management extends to people management activities in the organization.
7. Conflict resolution – Conflict is inbuilt in the organization. People working in the same organization would have different standpoints and interests even while working for the same objective of bringing profits and prosperity to the organization. Quality department would like to take maximum amount of time possible to detect every single error while the delivery department would like the quality department to take the least amount of time to be able to deliver before or on time; marketing department would like shorter delivery times and delivery department would like to have more time to build a robust and defect-free deliverable; finance department would like higher prices to show more profits while marketing department would like lower prices to beat the competition. In this manner, conflict is inherent between different entities in the organization. The conflict originates at lower levels but quickly gets escalated to the level of the departmental heads. When it reaches this level, the onus is on the top management to resolve the conflict to ensure that the resolution percolates down to lower levels and dissipates the conflict at its roots.

Now let us take a look at the role and responsibility of the HR department.

HR department – Most people in the organization think that the HR department does everything connected with the people in the organization. Unfortunately, it is a fallacy. HR department implements what the top management sets as policies and takes ownership of the policies. Also, it is a bridge between the people in the organization and the top management. HR department takes ownership of the people in the organization and with the direction from the top management and the help of the line managers, ensures that the people management activities are carried out fairly and effectively in the organization. Here are the responsibilities entrusted to the HR department in organizations, normally:

1. Take ownership and champion the cause of fair treatment of people in the organization – Every department in the organization takes ownership of something concerning the organization. Marketing department takes ownership of the customers; finance department takes ownership of the money; delivery department takes ownership of the products and services; similarly, HR department takes ownership of the people working in the organization. By taking ownership, HR department handles

all aspects concerned with the people in the organization ranging from recruitment, induction, internalization, promotion, grievance resolution to giving a sendoff while leaving the organization and settling the final payments.

2. Take ownership and champion the definition and improvement of the processes related to people management in the organization – In the present day, organizations are not run by the autocratic means of person-dependent style of management. It is rather being carried out in a process-driven manner. While the organizational process definition and improvement are coordinated by the organizational process group, each department takes ownership of the processes concerning it. Similarly, HR department, in coordination with the organizational process group initially defines the people-oriented processes in the organization and then, coordinates the process improvement adhering to the process defined for process improvement in the organization. It analyzes the process improvement requests sent in by various agencies in the organization and shortlists those that are likely to be beneficial for the overall good of the organization. Then in coordination with the organizational process improvement group, implements those changes in the processes and after reviews and approvals, rolls them out for implementation and internalization.

3. Diligently implement and internalize the people management processes in the organization – HR department implements the approved processes within the department as well as in the organization and troubleshooting any initial troubles and then works toward internalizing them in the organization. This activity is also the responsibility of the HR department not only during the initial stages but also when the processes are improved and approved for implementation.

4. Provide the required human resources to the organization at the required time, in the needed quantity, at the right cost and at the right quality – This is the recruitment activity and is one of the prime responsibilities of the HR department. It is discussed in detail in Chapter 5.

5. Ensure that the people in the organization adhere to the defined processes diligently – This is the responsibility of the HR department and it is performed as part of the organizational audits in which any deviation is brought out as a non-conformance. HR department takes two kinds of actions, namely, the corrective actions and the preventive actions. Corrective actions, as the name implies, correct the immediate

non-conformance and the preventive actions, ensure that this kind of non-conformance would not recur in future.

Now, the roles of the HR department in the people management activity are:

1. Take all actions necessary to implement and internalize the defined processes for people management in the organization – The people working in the organization have to internalize quite a few processes in addition to those processes that govern their work. One of them is the HR process. HR department needs to continuously train, or coach, or counsel the people on the HR process to clarify their doubts or in the way they need to be interpreted or implemented and so on. This will be on the basis of need as and when clarifications are sought by the people in the organization.

2. Continuously monitor the implementation of the defined people management processes and uncover opportunities for improvement – As in any other case, the people management policies come into question and improvements, which can even be changes, could be asked of the processes. HR department needs to clarify and convince or take a process improvement request and implement them in the processes. It also needs to take notice of the non-conformance reports being raised in the audits and take necessary actions to see the level and quality of the process implementation in the organization. Where necessary, HR department needs to intervene and correct the implementation in coordination with the head of the departments and the organizational process group.

3. Receive, analyze and shortlist all process improvement suggestions and take all actions necessary to improve the implemented processes in the organization

4. Carry out all activities efficiently and ensure that the organization does not suffer from shortage of the required human resources. – It is a fact that people leave the organization, not at the convenience of the organization but at their own convenience. Therefore, organization faces a crisis when people leave at an inconvenient time especially, when the people leaving are critical for the project success. HR department needs to take appropriate action to ensure that no project suffers due to some people leaving the organization. This aspect was dealt in greater detail in Chapter 5.

5. Perform all people management activities entrusted to HR department on time, effectively and fairly – "For want of a nail, the battle is lost" or

so goes the adage. Time is the essence behind success. Timely actions ensure success and delayed actions may cause failures. Remember the adage, "a stitch in time saves nine"? HR department needs to perform all its actions on the specified and agreed timelines to ensure organizational success.

6. Control costs – As the HR department is a cost center, cost control is necessary for the organization to be more profitable. HR department needs to continuously scan the environment to automate its tasks and minimize its costs. We discussed some metrics in the Chapter 13 and HR department uses these and other metrics specific to the organizational environment to reduce its costs on a continuous basis.

7. Skill retention and development – HR department needs to periodically evaluate the skills of the people working in the organization and discover the skill gaps with the assistance of the technical departments and take all necessary actions to bridge the gaps to keep the organization at the state-of-the-art in the chosen field of the organizational endeavor. This topic was dealt in greater detail in the Chapter 11.

8. HR department would ensure that all the payments due to employees are paid on time. These include the salaries, wages, bonuses, performance incentives, and other perks on time without causing any inconvenience to the employees. While making these payments on time may not lead to additional motivation but delays can cause demotivation. This topic was discussed in Chapter 8.

9. Performance management, motivation and morale – These topics are related and are predominantly in the bag of the HR department. HR department needs to monitor these aspects of all the people working in the organization periodically and take all necessary actions to maintain them at the desired level. These topics were discussed in detail in Chapters 9 and 10.

Now, let us discuss the responsibility and the role of the organizational process definition and improvement group.

Process definition and improvement group – Most of the professional organizations of today are driven by a well-defined and continuously improving set of processes that govern how each entity in the organization carries out its work and produces the deliverables expected of it. All such organizations would have a department which mainly consists of a few process experts that take ownership of the organizational process assets and continuously improve upon them adhering to a defined process for process improvement. That core group of process professionals draw upon the other organizational professionals on a temporary basis to actually define the processes initially and then implement the approved process improvement requests. Now let us see what responsibilities ought to be entrusted to the organizational process group:

1. Take ownership of the organizational process assets including initial definition and continuous improvement – Organizational process group is the central agency in the organization for any issue in regard to the organizational processes. They will coordinate the initial definition of the processes. In this work, they will take assistance of the experts from the concerned departments and in our case, from the HR department. Once defined and approved, the processes would be released for implementation. During the implementation period, the process group would handhold all stakeholders and troubleshoot any teething problems and ensure that the processes are smoothly implemented across the organization. Once implemented, the stakeholders are likely to raise process improvement requests to handle the changes in the technical and business environments. The process group is the agency that receives all such process improvement requests and consolidate them. Periodically, the process group arranges their analysis using the experts in the concerned departments and in our case, the HR department. The HR department selects some or all the process improvement requests for implementation. Then the organizational process group coordinates the work of modifying the organizational processes to implement the selected process improvement requests taking assistance from the HR department. Once the processes are updated, process group will carry out all actions to obtain the approval of the top management and release them across the organization for implementation. All in all, organizational process group coordinates the process definition, maintenance and improvement of all the organizational processes in the organization including those of the HR department.

2. Champion the internalization of the approved processes in the organization – Some people in the organization are averse to the process-driven working, especially during the initial days, as it brings discipline in the organizational working as well as checks and balances including quality assurance. Such people perceive process-driven working as shackles on their freedom. This is true to some extent as processes do not

allow unfettered freedom but it allows waivers in deserving cases. So, the organizational group ensures that all such opposition to the process-driven working is dissipated and that all people in the organization adopt the processes wholeheartedly.

3. Maintain the process assets library and make it available to all in the organization – Once defined, an organization would have many process documents that needs careful storage. Of course, all these documents are in soft copy form these days but somebody ought to ensure that unapproved documents are not used as reference. Organizational process group is the one entrusted with the responsibility of maintaining the organizational process documents such that no one in the organization uses an unapproved document for any organizational activity. This, they would achieve by a system of tight configuration management of the process documents with strict system of approvals for checking in the updated documents. All such documents are available to anyone to read and implement them in the work but not to modify those documents. Configuration management is usually carried out by the organizational systems administration department or such other department which maintains the organizational computer assets including servers. Organizational process group provides the authorizations and the actual work is carried out by the systems administration department.

Now let us look at the role played by the organizational process group in the management of people working in the organization –

1. Initiate and assist the HR department in defining the people management processes in the organization – While the organizational process group owns the organizational process assets, it is for the HR department to do the actual work of process definition and modifying those processes for improvement. While so, it is the process group that always takes the initiative in both the cases of initial definition and continuous improvement periodically. It initiates the initial definition and also when some process improvement requests are received. Process group coordinates entire effort from assigning the work, actual definition or modification for improvement to reviews and approvals for implementation and then the actual rollout and implementation. Thus, organizational process group assists in the definition, improvement, approval, rollout and internalization of people management processes across the organization. Especially the process improvement part is a continuous activity.

2. Assist the HR department in the implementation and internalization of people management processes in the organization – All people related processes are implemented by the HR department across the organization. The activities of actual release, then maintain these processes in the organizational repository and then handholding the employees during the initial phases of implementation rollout and the rollout after they are improved would be looked after by the organizational process group. While HR department develops the processes, the process groups performs all other activities connected with process implementation in the organization. Thus, it renders assistance to HR department during the implementation phase.

3. On behalf of HR department, receive all process improvement requests raised by all the stakeholders, consolidate them, arrange for their analysis, shortlisting and coordinate their implementation in the organizational processes – Process improvement requests would emerge not only from the HR department but also other employees in the organization when the business and technical environment changes take place. The organizational process group as the champion of the organizational processes, receives all process improvement requests including those of the HR department. While the process improvement can be raised at any time, the organization cannot implement it as soon as it is received. That way, the process assets would never be stable and there would be a process rollout frequently at short intervals that would inconvenience the employees. Therefore, the industry practice is to effect process improvement once in six months generally. This coincides, usually, with the organization-wide process implementation or certification audit which enables discovering opportunities for improvement during the audits. But there would be exceptions and if there is an emergency that needs immediate implementation, it would be implemented. So, the process improvement group receives all the process improvement requests being raised across the organization and performs a preliminary analysis and assigns a priority for implementation to each request. Then it categorizes each request by the basis of the department to which it is concerned. Then it consolidates the process improvement requests department-wise and on the specified date, these are handed over for analysis and implementation by the concerned departments. Then when the concerned departments including the HR department performs analysis and shortlists the process improvement requests for implementation which will

then be approved and taken up for implementation by the process group. Thus, the organizational process group assists HR department, along with the other departments of the organization.

We will be discussing the organizational process group and its activities in detail in the coming Chapter 15. Having discussed the role of the organizational process group in the people management activities, now let us look at the responsibilities and the role of the supervisors and managers in the organization in the matter of managing people at work.

All supervisors and managers – All employees in the organization report to someone placed higher in the organization. The superior is generally referred to as a supervisor or a manager depending on the culture of the organization. The actual work carried by these people is supervising the work of their subordinates in addition to the work they need to personally carry out. Anyone in the organization to whom some other employees report is to be referred to as a supervisor, irrespective of whether his/her designation includes the word "supervisor". In reality, it is these supervisors that carry out the actual activities of people management in the organization. True, they make use of the organizational framework put in place by the top management, the organizational processes put in place by the organizational process group in their work and the support offered by the HR department but, it is the supervisors that actually manage the people in the organization. These people have a vital role in the effective management of the people. The aspect of work management was dealt with in the Chapter 6. Let us now enumerate the responsibilities of the supervisors in the people management activities:

1. Take ownership the people reporting to them – People spend a third of their day and at least half of their wakeful time in the workplaces. So, the supervisors and managers become a sort of family heads to the employees. Just as a head of the family takes care of the family members, the supervisors and managers need to take ownership of their subordinates. The supervisors and managers have a dual responsibility, one toward the organization and the second to their subordinates and both interests must be protected. Taking ownership includes, inducting the employee into the team, getting them to perform their assignments effectively and efficiently, measure their performance, rewarding them for both positive and negative performance, motivating them, promoting them, and so on until either of them leaves the organization. Of course, the managers need to take assistance of the HR department in

all these responsibilities. If HR department recruits, it is at the behest of supervisors and managers; if HR department promotes someone, it is at the recommendation of the supervisors and managers and if HR department discharges someone, again, it is at the order of the supervisors and managers. While the ownership vested with the HR department is symbolic, the ownership of the organizational human resources by the supervisors and managers is real.

2. Manage the entrusted people to maximize/optimize their performance to further the organizational goals – This is perhaps the most important responsibility of the supervisors and managers. People are entrusted mainly to get some organizational work done by the people and when people perform, they not only need to earn their salary but also some profit to the organization. It is the responsibility of the supervisors and managers to see that the employees add value to the organization by their presence and work. Of course, the supervisors and managers take the assistance from the HR department, industrial engineers, behavioral scientists and so on but the primary responsibility is owned by the supervisors and managers.

3. Implement the approved people management processes diligently – In these days of process-driven professional organizations, it is imperative that each organization has a set of processes for the people to adhere to in their working. It is the primary responsibility of the supervisors and managers to ensure the implementation of all the approved organizational processes both in letter and spirit. They can always look for opportunities for process improvement and raise requests for the same but until a new process is approved and released, the present process must be adhered to and it is the responsibility of the supervisors and managers to ensure that the process implementation is carried out diligently.

4. Ensure that all the people in his/her charge are motivated and the team morale is maintained at the levels specified by the organization. While some of the motivating factors, especially the hygiene factors are in the purview of the organization, the motivating factors are in the purview of the supervisors and managers. The aspects of motivation and morale are discussed in greater detail in Chapter 9. It is well accepted that while the hygiene factors the lack of which can demotivate the people, their presence alone cannot achieve higher morale or motivation. The motivating factors among which are fair treatment at work and good supervision, can be ensured only by the supervisors and managers.

Perhaps, this responsibility is the most important responsibility of the supervisors and managers. Sometimes, there can be situations that are beyond the supervisor/manager that adversely impact the motivation, then, the onus is on the supervisor/manager to call for the assistance of the HR department for retrieving the situation to put motivation and morale on the right path.

5. Raise process improvement requests on the organizational process groups as necessary – Organizational process group might have defined the best processes, but we have to note that there is no absolute "best" in the world! When the environment changes, the technology changes or some other factors like the statutes change, we need to improve the processes to keep them current and effective. Being close to the revenue earning people, the supervisor/manager is best positioned to discover opportunities for process improvement. Therefore, they ought to raise a process improvement request whenever they discovery an opportunity for improvement. Only when there are such requests, the process group can initiate action for modification of the existing process and eventually upgrade the process to meet the challenges of the changed environment.

6. Cooperate with the HR department and the process group on all the people management initiatives – On behalf of the organization, HR department would launch many initiatives to improve the motivation and morale in the organization. These include various award and reward schemes, feel-good activities like the picnics, team outings, various celebrations like the annual day and so on. If they have to be successful and achieve their set goal, the whole-hearted cooperation of the supervisor/manager is sorely needed. Without their cooperation, every such initiative will come a cropper. Who else would benefit the most if such initiatives are successful besides the supervisor/manager?

Now let us look at the role played by the supervisors and managers in the organization in the matter of management of people at work.

1. Allot work fairly to all the subordinates so that each individual has comparable workload and that no person is either underloaded or overloaded with work and perform all activities described in Chapter 6 diligently. Work management is critical in the motivation and morale of people. That is why an entire chapter is dedicated to this topic.

2. Continuously scan the environment for opportunities for improvement in the aspect of managing people at work and suggest them to the

management and the HR department so that necessary action can be taken to improve the situation.

3. Whenever a grievance is raised by any of the subordinate, be the first level of grievance resolution. No one in the organization knows the grievance situation as thoroughly as the immediate supervisor/manager besides the person raising the grievance. This is because, the supervisor/manager has the first-hand information. All others take information either from the person or the supervisor and therefore, would not have first-hand information. Supervisor needs to use his/her good offices and try to resolve it by coordinating with the HR department and others as necessary. Only when it fails, should the supervisor escalate it to the next level. This has the twin benefits of (1) the resolution is the quickest and (2) the supervisor besides the organization gets the goodwill of the person whose grievance was resolved. Both these are beneficial to the organization.

4. The supervisor is the first person in the matter of performance measurement. All others next in the chain can only improve this measurement using their experience. Therefore, it is of paramount importance that the performance measurement be performed accurately and diligently. The general barriers to accurate performance measurement like the impact of recent incident, a dominating incident, one-time super/poor performance etc. should not cloud the measurement. The supervisors, ought to disengage their minds from all such influences while carrying out the performance measurement. This aspect has been discussed in the Chapter 10.

5. In the aspect of skill retention and development, again, it is the supervisors who can identify the gaps in the skills of their subordinates at the earliest. The need for enhancing a person's skills is noticed first by none other than the supervisor as the person works closest with the supervisor. Supervisors also would know first when the technology upgrade initiative is launched by the organization and has to take the lead in assessing the skill-gap existing in his/her subordinates. Therefore, supervisor needs to initiate the action early enough to upgrade the skills of the subordinates so that they are ready with the needed skills when the technology is eventually upgraded by the organization. The supervisor needs to interact with the HR department to choose the curriculum for the training program, the method of imparting the needed skills, assist in conducting the training programs, and then finally, give feedback

on the training program's effectiveness in the skill upgradation of the participants.

6. Every organization needs to groom its people for shouldering higher responsibilities. It is always better to promote internal people than recruit outsiders, most of the time. Sometimes we need to infuse fresh blood and viewpoint into the organization but most of the time, it would be advantageous to promote from within because their loyalty to the organization is known. So, it is the managers who can spot talent that can shoulder higher responsibilities. Most people have multiple talents and when they work in the organizations, only a limited set of their talents are utilized. The organization can fully utilize the talents of its employees, once their talents are known and managers can play a major role in learning and documenting the relevant talents of their subordinates to enable organizations to make use of those talents as and when necessary.

Now, having discussed the roles and responsibilities of the supervisor/ manager, let us now discuss the roles and responsibilities of the normal employees.

Employees themselves – In the professional organization, every one is an employee! So, why should we discuss their role here? We include in this category, those employees that are neither in the top management echelons nor have adequate number in their team to be called a supervisor or a manager. These employees do not define a process or set a policy. They work inside the framework set in place by the top management, using the processes put in place by the organizational process group and deliver the deliverables expected from them. Now, let us discuss their responsibilities and roles in the matter of managing people at work.

Wait a minute, are these the people about whose management we are discussing throughout this book? Yes. But, the "managed" people also do have some responsibilities! Here are their responsibilities –

1. Adhere to the approved processes in their working – The most important responsibility for these people is to adhere to the defined and approved processes released in the organization. We should give every opportunity for the approved processes to succeed even when we do not like them. Once a process does not beget the results it was supposed to, the management itself would modify them. Non-implementation of a process is killing it in its birth and we ought not to do that. If, necessary, we may take waivers to objectionable portions of the process but the rest needs to

be implemented. After all, we need to remember that the processes were developed to achieve organizational goals and before they were released, they were subjected to quality control activities like peer review and managerial review. If too many waivers were asked, the process group itself will review the process once again and initiate its improvement. So, it is always better to implement the process wholeheartedly and ask for changes than avoiding it.

2. Raise a process improvement request whenever they see an opportunity for improvement – Just because a process was approved for implementation in the organization does not mean that it was engraved on stone making it impossible to change! Processes like most man-made things can be changed. So, we should continuously scan our processes for opportunities for improvement. However, the improvements should not be to decrease our work by increasing it for someone else. The improvement suggested ought to be such that it improves either the deliverable quality or reduce the tedium or improve the team productivity. It needs to bring about all-round benefit. Whenever we find an opportunity for improvement, we need to raise a process improvement request. Sometimes, we may not be authorized to raise such requests and in such cases, we need to bring it to our supervisor/manager to raise the request.

3. Raise an issue whenever they find that either the hygiene factors are motivating factors become conspicuous by their absence. This ought to be done following the channels approved for use in the organization including the suggestion boxes which allow anonymity. Motivation of the people and the morale of the team is not only for the benefit of the organization. We individuals also need it. If we work in a team whose morale is low, our self-esteem itself takes a hit. In order to keep up our own self-esteem, we need to ensure that all others working in our team have high motivation levels. The fall in the motivation levels can have several reasons. We may not be in a position to pinpoint the reason accurately. But we can raise an issue with the appropriate authority, so the concerned executives can take necessary corrective and preventive actions. We ought not to shirk raising an issue when we are faced with one. By doing so, we not only help ourselves but also the organization.

4. Raise grievances whenever there is some genuine grievance – It is common for people in large organizations to have some grievances. When bulk data is processed either manually or by automatic means, some errors do occur. Because of these errors, some persons can be

aggrieved. This results in a grievance. If these grievances are left alone, they degenerate into frustration and passive opposition and then to open non-cooperation. That is why, every organization would have a grievance resolution mechanism about which we discussed in Chapter 7. But that mechanism would swing into action only when somebody raises a grievance. Some people are afraid to raise a grievance, even when they have a genuine grievance, fearing reprisals. But unless, a grievance is raised, the concerned executives would not even know if a mistake was made or a computer program is committing errors. In order to prevent errors in future, we need to raise a grievance when we have one. This action would not only resolve our grievance, but it also has the capability to prevent future errors and thus prevent grievances.

Now having discussed the responsibility of the common employees that do not have significant number of subordinates, let us turn our attention to their role in the management of people at work. Their role is to perform the work for which they were recruited diligently to their best capacity and cooperate with the concerned departments, managers and supervisors in all the organizational initiatives to improve and maintain the people management practices at their desired levels.

15

Process Definition and Improvement for People Management

15.1 Introduction

In the earlier days when the classical texts on personnel management were written, the organizations were managed based on the intuition and hunches of the managers. If something undesirable occurred, the concerned individual was let go. Every manager (or for that matter every individual) in the organization was accountable for his/her actions. This made organizations person-dependent for their success or failure. Large organizations realized that this over dependence on the individual is not a good augury for the long-term success and survival of the organization. We can punish the erring individual but the consequences of the wrong action are still on our hands. The organizations realized that prevention of errors is much more valuable than punishing the individuals for their mistakes. This brought in the culture of slowly eliminating the person dependency in the organization by the definition of standards, guidelines, formats, templates, checklists, and procedures. They brought in systems. They slowly but surely drove the organizations to be independent of the persons working in the organizations.

The person-dependent management produced some heroes/heroines who are worshipped even today. Many organizations reverted to their pits once the hero/heroin departed. The introduction of ISO 9000 series of standards changed this perception. These standards focused on process-driven working in the organizations. ISO (International Organization for Standardization) also formalized mechanisms to audit the organizations for compliance to their 9000 series of standards and certify them as compliant to ISO 9000 standards. Large organizations placing large procurement orders began insisting on the ISO 9000 certification from the vendors. This changed the industry to look at process-driven working and convert their organization from a person-driven organization to a process-driven organization.

In 1998, SEI (Software Engineering Institute) of Carnegie Melon University released their CMM (Capability Maturity Model) for software development organizations, which was adopted by many software development companies mainly due to the principals insisting on it. Many CMMs have emerged. Some became popular and some did not.

But these initiatives have changed the perception of many organizations including the small, medium and large organizations to adopt a process-driven management of the organizations. Now, processes are here to drive the organization and they would stay here for the foreseeable future. So, the managers, either at the top level or at the middle level, have to deal with it, adapt to it and work with in it.

15.2 Process

Before we talk of process improvement, we need to understand the process itself. When we attempt to produce and deliver something, we need to follow a series of steps in order to have the final deliverable. In our life everything follows a process, even if it is not explicitly defined or documented. In some areas, the process needs to be strictly implemented. For example, in cooking, the process is very strict and any liberty taken with the process is likely to result in poor taste or over cooked or under cooked food. In certain cases, we are allowed some freedom to skip some steps or poorly perform a few steps and still the consequences would not be readily apparent. For example, if we do not prepare the surface properly for painting, the paint would look alright but it would peel off quickly later on.

When we come to organizational processes, the importance of process definition and adherence was first recognized in process industries such as chemical manufacturing including fertilizers, drugs, petroleum products, plastics and so on. Most of these manufacturing was carried out by the machines. The human beings ensured that the process is adhered to and effect any required corrections when any deviation to process adherence was noticed. Any deviation above/below permissible limits had disastrous consequences by producing unusable products. Not so significant an importance was given to process compliance in discreet manufacturing except in areas like heat treatment, vacuum creation (in light bulbs etc.), semiconductor components such as transistors, IC chips and such other processes. Process importance and adherence became paramount when correction of the product is not possible afterwards as in the case of process industries. Where rectification of the defective product was possible, process did not gain that much significance as in the case of discreet manufacturing.

However, it was found that definition and documentation of process gives benefits to the organization, namely,

1. It gives detailed insight, for the top management, into how the work is carried out in the organization.
2. It facilitates analysis of the process by experts in the field. Their analysis helped plugging the loopholes in the process as well as improving the quality and productivity.
3. It facilitated more credible planning and scheduling of the work.
4. It facilitated uncovering the opportunities for automation/improvement.
5. It facilitated development of standards for methods of working as well as components.

Realizing these benefits, the discreet manufacturing industry began using defined process. This coincided with the development of the branch of Industrial Engineering comprising of the work-study principles and techniques.

It came to be accepted that definition and documentation of process improves quality and productivity. Initially the process orientation was called as "good manufacturing practices (GMP)" and this term continues to be used to this day.

The ISO 9000 series of process standards rather provided the much needed "push" to the industry to embrace process-driven working.

15.3 The Steps in Implementing a Process in an Organization

Now the steps in achieving process driven working in an organization are –

1. Process definition
2. Process implementation
3. Process stabilization
4. Process improvement

15.3.1 Process Definition

The first step in process definition is to assign the responsibility for championing the process definition and improvement to a department in the organization. Some organizations assign this responsibility to quality assurance department and some organizations assign it to a specialist process group. Whosoever is entrusted with the responsibility of championing the process-driven working in the organization would carry out process definition, initially and later on receive suggestions for improvement, and implement qualified suggestions in the processes after evaluating their benefits and concomitant costs. The actual definition of each of the processes would be carried out by the practitioners in the organization with the facilitation provided by the process group.

There are two approaches to defining the process, namely,

1. Top-down approach – this is suitable when the organization is new and the processes are being set up.
2. Bottom-up approach – this is suitable when the organization is already in existence and operations are being performed for some time.

Top-down approach to process definition – In this we perform the following steps –

1. Breakdown the organizational operations in a functional manner. The first level of breakdown would consist of major activities of organizational operations – that is we breakdown manufacturing into marketing, production, quality assurance, finance, HR, and so on at the first level. Then each of these major activities on the first level is further broken down to their next levels, such as the HR department is broken down into recruitment, employee relations, skill retention and development and so on. We continue the breakdown until we feel that further breakdown does not add any further value.
2. For the each of the activities in the first level of breakdown, we define a process.
3. We define a sub-process for the next level, if it consists of sub-levels.
4. We define a procedure for an activity that is not broken down into further levels.
5. Wherever it is possible, we define standards and guidelines to aid practitioners in adhering to the procedures.
6. We define formats and templates for the purpose of capturing, recording and presenting information to achieve uniformity among different practitioners in the organization.
7. We define checklists to aid practitioners in adhering to the procedures and performing the activities comprehensively and exhaustively.
8. We define measurement and analysis of the results of the process to evaluate the efficacy of the defined process.
9. We arrange for review of the defined processes by practitioners in the organization to ensure that the defined processes reflect the reality and that the process is accurately defined.

Bottom-up approach to process definition – In this approach we perform the below steps –

1. We study how the practitioners are performing their tasks and document them. It is obvious that there would be differences among persons, even within the same organization, in the way they perform their activities. Therefore, we take the most common practice as well as the practice of the persons that are known to produce the best quality deliverables.
2. We also capture the formats and templates being used by the practitioners. We study them and develop new formats and templates capturing the best features of different formats and templates being used in the organization.
3. We capture, from middle managers and top managers, the details of the processes and document them, culling the best practices from all managers.
4. We organize the collated material into processes, procedures, formats, checklists, standards and guidelines.
5. We release a draft version of organizational process and invite comments from all concerned persons in the organization.
6. The feedback received would be analyzed and all feedback that would either enhance quality or productivity or simplify the work would be implemented in the process. Then we release the process for implementation in the organization.

15.3.1.1 Building quality into the defined process

Once a process document is prepared, before releasing it for implementation, we subject it to scrutiny by experts in the field either from within the organization or from outside the organization. These experts would evaluate each component of the process and compare it with the best practices in the industry. This evaluation and comparison with the best practices would produce a gaps analysis document. Now the in-house process group would analyze each of the gaps uncovered for its implementability in the organizational process and practice. This may necessitate additional investment in tools, training for personnel and change in the methods of working and so on. Sometimes it may be possible to implement the best practice fully and sometimes, it may be tailored to suit the organizational environment for implementation. Sometimes, we may have to totally reject the best practice, if it does not suit the organizational environment. In line with the decision on implementing the suggested best practices, the process documents would be updated and finalized.

This inclusion of best practices into process documentation first and then putting them into practice next would ensure that quality is built into the process.

Then we include the quality assurance activities necessary, into the process documents to ensure that the defined process is subjected to quality assurance activities during practice. These may include reviews, tests, inspections and audits. Then we also include the measures and metrics necessary to assess the performance of the process.

This would build quality into the process.

15.3.1.2 Aligning the process with a process model such as ISO 9000/CMM

Now, the requirements of aligning the organizational process to a process model such as ISO 9000 or CMM is becoming essential as more and more customers are insisting on a certificate of compliance or a maturity rating as a prerequisite for participating in the bidding process for purchase contracts. However, if we define our organizational process comprehensively and including the industry best practices, it would be adequate to meet any model, because the main goal of all models is to ensure that the organization utilizes industry best practices in its functioning and producing the deliverables. Once we ensured that the organizational process is utilizing the industry best practices, all that needs to be done is to confirm that the selected model and our organizational process are in sync with each other and that our process meets or exceeds the selected model's goals.

In order to align the organizational process with such a model the following steps would be necessary:

1. Study the model requirements especially the goals that are to be achieved by the organizational process and practice.
2. Carry out a gap analysis between the model requirements and the organizational process. These gaps could be overshooting gaps, that is our practice exceeds the requirements of the model or shortfall gaps, that is our process does not meet the model requirements.
3. Enumerate all shortfall gaps, that is, the instances in our process where our process does not fulfill the model's goals, into a gap analysis document.
4. Carry out a series of consultations with the organizational practitioners and management about the gaps and the ways to bridge them.
5. Select the most suitable alternative solution to bridge the gaps and implement them in the process.

6. Try out the improved processes on a pilot basis on a few projects.
7. Implement the relevant feedback from pilot implementation in the processes.
8. Arrange for review by practitioners from the organization.
9. Release the new set of processes for implementation in the organization.

These steps ensure alignment of organizational processes and practices with those of the selected process model.

15.3.2 Process Implementation

1. Define the process
2. Carry out quality assurance activities
3. Obtain approvals
4. Implement the process on a pilot basis
5. Obtain the feedback and implement it in the process
6. Obtain approvals
7. Conduct training on implementation
8. Implement the process across the organization

   (a) Big bang approach
   (b) Phased approach

9. Handhold
10. Move into improvement mode

These steps would ensure that the process is implemented diligently in the organization.

15.3.3 Process Stabilization

Process stabilization is necessary for an organization to produce predictable results. That does not however, mean that process improvement is not necessary. It only means that the process ought to be by and large a stable one and improvements are effected based on a trigger and such improvements would be to improve quality or productivity or simplify the work. Process stabilization becomes possible only after implementing the process. An organization typically goes through these stages before having a stabilized process –

1. **Initial stage** – this is when the organization came into being and started its operations. It is trying to establish commercial viability. Operations are performed based on personal direction of the owners/chief executive/senior management.

2. **Process definition** – Once the organization achieved commercial viability, it defines the processes for conducting operations using a process-driven approach.

3. **Process implementation** – the defined process is implemented in the organization. All operations are conducted based on the defined processes. The process would be institutionalized in the organization.

4. **Process maintenance** – once process is implemented, it would be monitored and improved as required, based on events or time triggers, performing the following steps:

   (a) **Analyze results of the operations for any variances** – some variances could be desirable that is, those variances are better in quality or productivity. Some variances could be undesirable that is, the defects are more or the productivity is diminished.

   (b) **Conduct root cause analysis for variances** – we conduct a root cause analysis for undesirable variances. Some of these variances could be due purely to chance causes. Even in the most tightly controlled and machine-controlled processes, random variances do occur. Some could be due to assignable causes. The analysis separates undesirable variances into these two, assignable and random, classes. For the undesirable variances, which are due to assignable causes, it would be analyzed if they were due to process defects or other defects. If they are process defects, then we effect improvement in the defined process.

   (c) **Improve the process for variances due to assignable causes** – For those undesirable variances, which are due to assignable causes attributable to defective process, we effect improvements in the process to plug the loopholes so that those variances would not recur.

   (d) **Pilot improved process** – the improved process would be implemented in a few projects to observe the efficacy of the improvements. If the results show that the improvements did not produce the desired improvement, steps b and c would be iterated until the desired improvements are realized from the pilot implementation.

   (e) **Implement improved process** – once the pilot implementation of improved process showed that the desired levels of improvements are realized, it would go through the normal process for implementing the process in the organization.

5. **Stable process** – when all the variances are due to random causes (chance errors), then we do have a stable process.

Once we have a stable process, we can make use of statistical quality control techniques such as control charts.

A well-defined and institutionalized process is a prerequisite for fostering a quality culture in the organization. An organizational process would have basically four types of processes, namely,

1. **The technical processes** – these processes define how the deliverable is built. These would typically comprise of processes for specification, design, and production/operations
2. **The quality assurance processes** – these processes define how the quality is built into the deliverable as well as ensuring that it is indeed built in. These would typically comprise of verification, validation, inspections, and audits.
3. **The management processes** – these processes define how all the other processes are managed. These would typically contain the project management process (including marketing), planning, quality management, work management, human resource management, stakeholder expectation management, and so on.
4. **The support processes** – these processes define how the technical processes are supported by the organization. These would comprise of finance process, HR process, sub-contractor management process, facilities management process and so on.

15.3.4 Process Improvement

Once we have a set of approved processes for the organization that are stabilized, we need to continuously monitor its performance. This monitoring is with respect to its actual performance vis-à-vis the desired performance. We need to do this as the organizational climate keeps changing. The organizational climate keeps changing because of:

1. The changes in technology bring in new paradigms for our deliverables.
2. The competition that uses better tools and techniques to deliver better quality deliverables at a cheaper price.
3. The governmental regulations about safety, usability, reliability etc. may make it imperative to deliver better quality.
4. Availability of new tools that may save money or improve quality or reduce cost.
5. Availability of new components or materials that improve quality, reliability, safety or security.

All these make it imperative for us to upgrade our processes to meet the changing environment and be competitive in the market. Therefore, we need to approach improvement of our organizational processes in a disciplined manner – that is by using a process driven approach.

We need to define a process for improving our organizational processes while defining the other organizational processes. This process for process improvement would comprise of the following topics –

1. **Triggers for process improvement** – the triggers could be event-based triggers or duration-based triggers. Event based trigger could be the audit report of an external auditor, or external appraiser or a report of organization-wide internal audit/appraisal. It could also be because of the procurement of a new tool that would have an impact on the process. Or it could be the release of a new standard or a new version of our adopted process model by a standards body that impacts our organizational operations. Duration based triggers could be like once a quarter/half-year/year or at the beginning of a new financial year and so on. These triggers are used to carry out an exercise to consolidate all the process improvement requests, analyze them and effect process improvement.

2. **Sources of opportunities for process improvement** – These could include internal sources such as suggestions from technical team members, managers, top management, or external sources such as suggestions from external auditor/appraiser or a standards body or process model owners and so on.

3. **Procedures for placing process improvement requests** – it would give the procedure stating how to place a process improvement request, its format/template, the information that needs to be included in the request, approvals required for placing the request and so on.

4. **Authorized persons for placing process improvement requests** – it details the types of requests that can be placed by various persons in the organization. For example, the marketing executive may place a process improvement request about product/process specifications or testing practice. It will be incongruent if a marketing executive raises a request for improving defect analysis procedure. That is, a person is best suited to raise a process improvement request on an area in which the person is working. These criteria would be enumerated in this section.

5. **Procedure for analyzing and accepting process improvement requests** – this section would describe the analysis that could be conducted on the process improvement requests that are received by the process improvement group. The agency that is authorized for approving or rejecting the process improvement requests would also be included in this section. In cases of rejection, the action that needs to be taken to communicate the rejection to the originator of the request would also be mentioned here.

6. **Procedure for implementing the process improvements and reviewing the improved process documents** – this section would describe how to select persons for implementing the improvement suggestions, the versioning norms for process documents, the reviews to be conducted for the improved process documents, and approval authorizations etc. would be mentioned here.

7. **Procedure for pilot implementation for obtaining field feedback** – once the process documents are improved and approved, they need to be implemented on a few projects for obtaining feedback from field. This section would describe such a procedure including selecting candidate projects, giving them waivers necessary for deviating from the currently approved processes, obtaining feedback from pilot implementations, collating and analyzing the field feedback and so on would be mentioned here.

8. **Implementing the feedback from the field and releasing the process** – once the feedback from the pilot implementation is obtained, it has to be implemented in the process. This section describes how to implement the field feedback, organize the review, obtain necessary approvals and release the improved process for implementation. It is the normal practice to effect only a few releases per year based on duration such as once a quarter/half-year or yearly.

There could be deviations in this process from organization to organization based on their unique set of circumstances. The above describes a typical process for process improvement.

15.4 Components of a Process

A **process** is an overall definition for a major organizational activity. A process is an inter-related network of procedures for performing an activity. It is the top-level document under which all other documents would be describing the details of the activity. A process is an assemblage of the following components –

1. **Procedures** – procedures are step-by-step instruction for performing a sub-activity of the process. Examples of procedure are, recruitment procedure, grievance resolution procedure, internal training program procedure, audit procedure, defect analysis procedure, and so on.
2. **Standards & guidelines** – These define the organizational selection from among a set of available alternatives for use in the organization so that the output would be uniform throughout the organization. While procedures bring in uniformity in performing an activity, standards and guidelines bring in uniformity in the deliverables. Examples of standards and guidelines are interview guidelines, training gaps analysis guidelines, faculty assessment guidelines, selection of faculty guidelines, defect analysis standards and so on.
3. **Formats and templates** – these facilitate a uniform way for capturing, recording and presentation of information. Examples for formats and templates are leave application form, grievance reporting form, training needs analysis template, recruitment request template, process improvement request template, and so on.
4. **Checklists** – these assist the person performing an activity exhaustively and a reviewer in ensuring the comprehensiveness of the review. The checklist would contain a number of items against which there would be a place to note "yes/no/not applicable". As each point is completed or reviewed, we can check off yes/no/not applicable against that point. When we complete the activity, we can go through the list to ensure that all points are taken care of.

All these components would be assembled together, usually, in a software tool on the Intranet of the organization for all the concerned people to refer anytime they needed to. All these are referred to as the organizational process assets. All these artifacts ought to be developed in such a way that people can easily read them, understand them, assimilate them, implement them and internalize the process in the organization.

15.5 Process Certification

In the present day, certification or obtaining a rating from a model owner has assumed significance as many of the outsourcers of high value contract work are using this certification/rating to shortlist their prospective vendors.

Therefore, a chapter on process improvement cannot be complete without touching in this topic.

There are various models and certifications depending on the type of the industry. But **ISO 9000** of International organization for standardization is common and most popular among the models. A certificate of compliance for these standards is awarded by an authorized lead auditor. ISO 9000 standards have undergone revisions since release. Authorized lead auditor conducts an audit of a set of sample implementation in all departments of the organization for process compliance including the support groups for their process compliance and raises NCRs (Non-Conformance Reports) wherever the practice is not conforming to applicable ISO 9000 series of standards. Authorized lead auditor would grant the certificate of compliance if there were no NCRs, or the NCRs were not very significant. If the NCRs are very significant, the authorized lead auditor may refuse certificate or withhold it till the NCRs are satisfactorily resolved by the organization. ISO 9000 standards are available for purchase from ISO.

The appraisal method and the final award slightly differ in different models. The common steps in attaining the award are –

1. Define your process.
2. Implement the process.
3. Improve the process based on the feedback obtained from the field after the process implementation.
4. Stabilize the process – this would include internalization and institution-alization of the process-driven working in the organization.
5. Now the organization is ready for certification/rating.

ISO grants a certificate of compliance and performs surveillance audit once in six months to ensure continued compliance.

Books are written on these models and abridging them is very difficult and covering the subject doing complete justice to those models is beyond the scope of this book. If you are interested in learning more about them, our humble suggestion is to obtain their documentation and study them. Alternatively, you can approach a process consultant and obtain the needed guidance.

15.5.1 Role of Top Management in Process Improvement

The first aspect of process improvement is the decision of the top man-agement to move the organization toward a process-driven management. This decision entails significant expenditure initially in defining a process appropriate for the organization; pilot it in sample locations; obtain feedback, analyze it and implement relevant feedback in the process; and then roll it out for organization wide implementation. The organization wide implementation would certainly disrupt the operations albeit in a limited manner for a limited amount of time. There would be stiff resistance from the managers to change from a person-driven model to a process-driven model as it takes away some of their freedoms. Then there would be expenses in defining the process, piloting the process, implementing the feedback, process improvement and so on. Then if a certification is desired, it would entail additional expenditure. Top management has to fund these resources. Then they have to ensure that all these activities are executed diligently.

Then the top management needs to demonstrate its commitment to the process-driven working. If not, the organization would quickly slide back into a person-driven model very quickly. Some organizations implement the process-driven model only for the certification and once the certification is obtained, they dump the process and revert back to the person-driven model.

If the process-driven model has to take root and give fruits, the top management commitment and demonstration thereof is essential. This is the top management's role in process improvement.

15.5.2 Role of Supervisors/Managers

As usual, the role of the supervisors/managers is to ensure that all activities of process definition, piloting, process roll out, process stabilization and process improvement are diligently carried out in the organization. However, not only the top management but also the supervisors/managers are also capable of sabotaging the process-driven model. Supervisors/managers need to work for the sake of the organization rising above their narrow political gains and victories in one-upmanship games. Once a process-driven model is implemented diligently, it decreases the chances of something going wrong. If something goes wrong, it enables to pinpoint why exactly it happened and gives a chance for taking corrective and preventive actions. If there is one section of organizational human resources that derives the greatest benefit, it is the supervisors/managers. So, it pays to have a process-driven model working well in the organization for the supervisors/managers.

The role of supervisors/managers is to carry out all activities of process definition and improvement as well as to work diligently towards achieving success to the process-driven model in the organization.

Index

About the Authors

Murali Chemuturi is a consultant on information technology and management, author and trainer. In 2001, he formed his own IT consulting and software development firm known as Chemuturi Consultants. Chemuturi Consultants help organizations achieve their management, process definition and improvement, quality and value objectives. The firm provides training in several software engineering and management topics. His firm also offers a number of products to aid project managers and software development professionals such as PMPal, a software project management tool; EstimatorPal, a software estimation tool and MRPPal, a comprehensive tool for material management.

Prior to starting his own firm, Murali is an industry veteran in various engineering, manufacturing and information technology management positions. He gained more than 49 years of information technology and software development experience. His final position prior to forming his firm was Vice President of Software Development at Vistaar e-Businesses Pvt., Ltd. In all, he worked in professional organizations for 31 years and as a consultant for the past 18 years.

Mr. Chemuturi's undergraduate degrees and diplomas are in Electrical and Industrial Engineering and he holds an MBA and a post-graduate diploma in Computer Methods & Programming. He has several years of academic experience teaching a variety of IT and management courses. The management courses he handles are operations management, management information and control systems, marketing and market research, productivity, materials management and project management besides soft skills.

351

Murali is a member of IEEE, a senior member of the Computer Society of India and a Fellow of the Indian Institute of Industrial Engineering, and he is a well-published author in professional journals. He had authored seven books on software estimation, software quality assurance, IT project management, software design, programming, requirements engineering and management. He co-authored one book on software project management with Thomas M. Cagley, Jr. He contributed chapters in four books.

Vijay Chemuturi graduated with honors from University of Scranton, USA, in 2003 with a Masters in Business Administration (MBA) with dual majors in Finance and Management Information Systems. He is a licensed Certified Public Accountant (CPA) in the state of New Jersey and is also a Certified Fraud Examiner (CFE). He has over 15 years of experience in a professional multi-billion dollar multinational organization. He is currently working as a senior manager in an industry-leader global public accounting firm. In his management career, he had led and managed the execution of audit engagements of large and multi-national Fortune 500 public and private organizations, development of Audit efficiency tools using Data Analytics and he currently oversees compliance with regulatory and firm policies in the risk management department. He published, along with Murali Chemuturi, a book on Management: A New Paradigm for the 21st Century in 2015. In 2011, he was invited to speak at United Nations in the World Forum on the Diaspora Economy – The Creative Economy. His paper, Are You Prepared To Assess Fraud Risk Factors? was published in the Pennsylvania CPA Journal. His definition of the term "fraud" is the most accepted in the industry circles and has been cited in research works and papers in journals.